Escape From System 1

ESCAPE FROM SYSTEM 1

SYSTEM 1

Unlocking the Science Behind
the New Way of Innovation

ANDREAS RAHARSO

Marshall Cavendish
Business

Published in 2022 by Marshall Cavendish Business
An imprint of Marshall Cavendish International

A member of the
Times Publishing Group

Other Marshall Cavendish Offices:
Marshall Cavendish Corporation, 800 Westchester Ave, Suite N-641, Rye Brook,
NY 10573, USA • Marshall Cavendish International (Thailand) Co Ltd, 253 Asoke,
16th Floor, Sukhumvit 21 Road, Klongtoey Nua, Wattana, Bangkok 10110, Thailand
• Marshall Cavendish (Malaysia) Sdn Bhd, Times Subang, Lot 46, Subang Hi-Tech
Industrial Park, Batu Tiga, 40000 Shah Alam, Selangor Darul Ehsan, Malaysia

Marshall Cavendish is a registered trademark of Times Publishing Limited

National Library Board, Singapore Cataloguing-in-Publication Data

Name(s): Raharso, Andreas.
Title: Escape from System 1 : unlocking the science behind the new way of innovation
/ Andreas Raharso.
Description: Singapore : Marshall Cavendish Business, 2022.
Identifier(s): ISBN 978-981-5009-31-6 (paperback)
Subject(s): LCSH: Creative ability in business. | Management.
Classification: DDC 658.4063--dc23

Printed in Singapore

To Gikie, Tania, and Fronia:
My three angels

Contents

Part III
Creating Next Practice from the Outside In

Acknowledgements

First and foremost, thanks to my undergraduate, graduate, and executive students from INSEAD Business School and NUS Business School. My book is infinitely richer for their contributions.

The following humans played a crucial role in making this book a reality.

My patrons: Beh Swan Gin, and Israel Berman.

My networks: Low Yen Ling, Yeoh Keat Chuan, Nick Walton, Wouter Van Wersch, Ralph Haupter, Martin Hayes, Harriet Green, Scott Beaumont, Damien Dhellemmes, Suran Suranjan, Ben King, Laurent Gatignol, Patrick de Moustier, Tricia Duran, Nutan Singapuri, Jon Ye, Leslie Hayward, Petrine Puah, Alvin Ng, and Tom Welchman.

My colleagues: Ruth Wageman, David Derain, Sylvano Damanik, Junichi Takinami, Madeline Dessing, Stephan Frettloehr, Renato Ferrari, Jeff Shiraki, Brian Langham,

Sia Siew Kien, Neo Boon Siong, Javier Gimeno, Felipe Monteiro, Quy Huy, Diny Sandy, Bridget Tee, Vivien Lim, Kulwant Singh, Ang Swee Hoon, Jochen Wirtz, and Andrew Delios.

My editors: Audra Lim, Justin Lau, and Melvin Neo.

Introduction

What if, as a human race, we are not as smart as we think we are?

We often have opinions about something, intuitions about people, know whether someone can be trusted – all without really being able to pinpoint how we know these things. We have answers to questions that we do not completely understand, relying on evidence that we can neither explain nor defend.[1]

In 2011, Nobel laureate Daniel Kahneman in his book *Thinking, Fast and Slow* noted that when faced with difficult questions, people simplify the hard task by substituting it with an easier question. It happens in a split-second, unintentionally, because of the mental shotgun our brains reach for instantaneously. It is easier to generate quick answers to difficult questions than it is to impose a cognitive load on our brains.

What is the meaning of happiness? Should I invest in Amazon? What are the likely political developments next year? Faced with these complex questions, our minds intuitively switch to answering questions such as what is my mood right now? Do I enjoy using Amazon's services? Do I like the current political candidate?

"The target question is the assessment you intend to produce. The (mental shotgun) question is the simpler question that you answer instead," explained Kahneman. Because mental shotguns make it easy to generate quick answers to difficult questions without imposing much hard work on our brain's lazy System 2 (more on this in a while), it means that we do not have precise control over our thought process and response. And we do not end up answering the original difficult question.

Sometimes this mental shotgun works well enough. Other times, it can lead to serious errors. Such as choosing a wrong candidate as president. Or approving an emergency medical procedure that is doomed to fail.

Our propensity to substitute hard questions with easier questions poses a big problem when it comes to innovation. Innovation is difficult. If we consistently gravitate towards easier questions, how will we ever be able to solve the difficult questions in innovation and create the new, and the best?

If Henry Ford did not think about how to create better modes of transportation and invent the Model T, we might still be stuck using horse carriages.

If Bill Gates did not have his bold and seemingly impossible dream to put a personal computer in every home, we might still be bound to using giant mainframes.

If Moderna did not think about creating a new vaccine technology to replace the old method of using inactivated viruses, the rampage of Covid would have created far more deadly consequences.

Only when we acknowledge that the process of innovation is incredibly arduous, and accept that we can use all the innovation frameworks in the world and still not have the secret formula to innovation creation, are we ready to harness a better way of innovation in our lives.

★　　★　　★

This is a book about the true nature of innovation.

Much of what we have come to believe about innovation has, with the latest in neuroscience and consciousness research, been shown to be wrong. Innovation is not something human beings are naturally able to do. It is not easy to stimulate creativity. Neither is it natural to make a fundamental shift in thinking from a place of default to a place of intentionality. Our brains are simply not wired that way.

Innovation is extremely difficult.

Evolutionary science has demonstrated that we instinctively seek efficiency, the path of least resistance, the tried and tested, to ensure that as a species, we survive. We do not deliberate long and hard about new ways to eat, cross the road, travel, function in daily life; we simply use the most efficient and commonsensical method that works.

In the same way, many companies do the same. The bottomline is key. Maximise profits, cut costs, be efficient, keep the business going. For as long as these aims are achieved, status quo is king. Just throw in a new product feature or update every now and then to keep the customers happy.

But that is no longer enough.

Great ideas that are unique and revolutionary are becoming increasingly difficult, and expensive, to mine and find. There has been debate in recent years about whether innovation is plateauing. Even as research efforts are on the rise, research productivity is on the decline.

Drawing on seven decades of scientific research in cognitive science, this book spans the works of three Nobel laureates – Herbert Simon, Daniel Kahneman and Richard Thaler. I will show you the mismatch between what science knows and what business does, and how that affects every aspect of innovation.

To do that, I am going to introduce Next Practice: an avant-garde course of action against critical challenges and problems faced by organisations that produces results superior to any Best Practices currently in use.

Next Practices are future-oriented, original, experimental and almost counter-intuitive. There is no existing benchmark simply because no other company would have done it before. Next Practice is essentially the first Best Practice: the forerunner of all best practices that follow within a field, sector, or organisation. Next Practices occur infrequently, but when they do, companies that do not recognise them because they are only prepared for best practices will fail to ride the wave. Once-great companies such as Kodak, Xerox and BlackBerry have fallen as a result of this blind spot.

Part I of this book looks at what Next Practice is and *why* it matters. Chapter 1 sets the scene with the rise of Next Practice in key industries of our economy, how it has helped companies to grow into S&P 500 companies by finding the shortest path to the innovation frontier. Chapter 2 will shake our beliefs about current innovation solutions, which are based – wrongly – on the assumption that human beings are born innovative. We are not, because we are trapped in what Daniel Kahneman calls System 1 thinking. Chapter 3 introduces the radiance of System 2 thinking, and demonstrates how vital it is as the creator for Next Practice work.

Part II shows you *how* to create Next Practice from the inside out, and primes your mind and organisation for getting to that sweet spot. Chapter 4 definitively points out the escape route from System 1, which is found in a hidden but prolific substance in our brains: myelin. This is followed by Chapter 5, where I reveal the vital importance of having a Quiet Brain for unlocking ourselves from System 1. Chapter 6 then brings in the unexpected use of failure as a weapon, and shows how we can conquer our fear in order to be bold and brave for Next Practice creation. Part II will end with Chapter 7, where you will apply training (T) and rituals (R) in conquering fear and myelin-building to strengthen the newly illuminated System 2.

Finally, Part III throws the spotlight on innovation frameworks that are trending, and how we can massively *level up* on them with Next Practice methodology. Chapter 8 demonstrates the application of Next Practice to Design Thinking and Working Backwards to make these approaches even more impactful. Chapter 9 wraps up with a deep dive into how Next Practice can enhance First Principles and Appreciative Inquiry – and supercharge them both.

★   ★   ★

**"The first rule of survival is clear:
Nothing is more dangerous than
yesterday's success."**
Alvin Toffler, businessman and futurist

We can no longer afford to work the way we have done through the last few hundred years. Those ways may have worked well, but they won't for much longer.

A paradigm shift is in order. Not just any shift, but a colossal seismic shift from what we have been used to.

Let us begin our journey.

PART I

What Next Practice Is and Why It Matters

CHAPTER ONE

The Rise of Next Practice

**"The real voyage of discovery consists
not in seeking new landscapes,
but in having new eyes."**
Marcel Proust

The data is sobering.

Companies are losing their tenure on the Standard & Poor's 500 list at a jaw-dropping rate. In 1958, the average lifespan of companies on the list was 61 years. Today, it is less than 18 years. Richard Foster, Senior Partner at McKinsey, further projected in his book *Creative Destruction* that by 2027, 75% of companies currently on the S&P 500 list will have disappeared.[2]

That is 3 in every 4 companies – merged, bought out or gone bankrupt by 2027.

While some of the oldest stalwarts on the New York Stock Exchange such as General Electric and ExxonMobil bear battle scars and retain their places, a fair number of companies with the largest market capitalisations are infants in comparison, having been in existence for barely a few decades. Companies such as Etsy, SpaceX, Amazon. Companies that come from sectors previously inconceivable: artificial intelligence, space, e-commerce. Some of these sectors have seen entire industries merge into 'hybrid industries' – notably digital healthcare, retail-tainment, and e-mobility.[3]

For traditional strongholds and start-ups alike, it is an era of disruption like no other. While disruptions are par for the course in business, it is the speed and complexity with which they are taking place, on a global scale, that is causing such an upheaval.

Business life expectancy, product life-cycles and relevance of past experience are all shortening. The reach and impact of new technologies, especially that of cloud, have created tailwinds which have either dramatically lifted or painfully crashed companies. Doing business has now evolved into a craft that requires a nimbleness to ensure survivability; a dexterity of thought, to be able to work with and around the disruptions.[4]

This nimbleness is crucial. Yet, it is elusive. Businesses often find it difficult to identify new directions to pivot towards, and to develop new products and services that will be the next big thing in wooing their customers.

New World Paradigm

The late Alvin Toffler, one of the world's most renowned futurists, wrote in his book *The Third Wave* (1980) that the waves of societal progress and evolution would get shorter, and faster. Societies would make dramatic transitions in shorter wavelets of technology, instead of massive and slow-moving grand waves. Using the Toffler Curve, he predicted that change would occur more rapidly, in vertical patterns, and accelerate, in turn accelerating the rate of disruption to humans and societies, resulting in a hyper-disrupted economy.

Turns out Toffler was right on every count.[5]

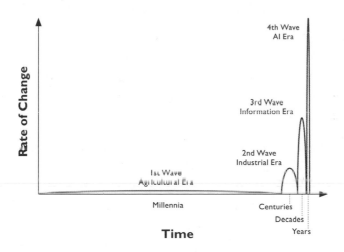

While it took thousands of years for man to go from the discovery of fire to the invention of the wheel during the Agricultural Age, and a few hundred years for the Industrial

Age to shape societies the way we have become familiar with, the rate of change in the present Information Age is measured in decades.

We are right at the cusp of the Artificial Intelligence (AI) Era, where change can occur in a matter of years or less. In fact, Buckminster Fuller's Knowledge Doubling Curve shows that knowledge took 500 years to double pre-Industrial Age, but the doubling started to speed up through the years, to a mere 12 hours in 2020. It is impossible for a human being to process all that information without the help of AI.[6]

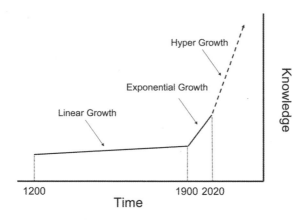

Discoveries and ideas no longer have time to naturally percolate as they used to. While they have not been depleted, ideas that are unique, original, and untapped are getting more expensive to find.[7]

Nicholas Bloom, economics professor at Stanford University, found in a 2017 research paper on economic growth and research productivity that when there were more researchers producing more ideas, it resulted in more economic growth. However, while research efforts were rising, the ideas being produced per researcher were declining sharply in number.

"It's getting harder and harder to make new ideas, and the economy is more or less compensating for that," Bloom said. "The only way we've been able to roughly maintain growth is to throw more and more scientists at it." The paper further states that the economy has to double its research efforts every 13 years just to maintain the same overall rate of economic growth.[8]

Bloom was in fact not the first to notice this element of intensification in processing ideas and information.

Economist Benjamin Jones noted in a 2009 paper that "if knowledge accumulates as technology advances, then successive generations of innovators may face an increasing educational burden".[9]

What does this mean for the researchers, innovators and entrepreneurs of our time?

Innovation depends on knowledge. Whether scientist or technologist, baker or barista, one needs access to a wealth of knowledge about one's industry, be it through direct or indirect means. Yet for an individual to know enough to do innovative knowledge work, it is very much akin to asking a professor whether he knows everything there is to know about the field he is in. There is simply way too much information that needs to be processed, in too short a time. Increasingly, an individual will only be able to master an ever-decreasing percentage of the total knowledge there is in existence.

This is known as the burden of knowledge. The more knowledge there is about a topic or field of study, the heavier the burden is on the subsequent person who tries to learn more about it. Data will keep getting updated. New research studies will keep being completed. It is almost futile to try and keep up.

And it gets even harder to beat the competition, find the next new innovation. Where the Best Practices have already been established, the burden of knowledge is immense. Even solutions such as forming teams to distribute the burden between members are not ideal – it gets expensive to hire more talent, while human dynamics can result in tensions that derail efforts.

Businesses today understand the dire need to innovate – or die. Yet the scramble to inject funds into R&D, make improvements to existing products, and tap on Big Data and customer analytics, barely makes a ripple in the busy and noisy marketplace of ideas.

What should businesses be doing then? Is it near impossible for businesses to make innovative breakthroughs before the intensification of the time-research ratio and burdens of knowledge engulf them?

Not quite.

Next Practice Solution

There is a way.

In my years as a business professor and consultant, studying countless case studies and collaborating with businesses in a plethora of industries, one thing always puzzled me.

It seemed as if businesses often benchmarked themselves against prevailing Best Practices, and tried to top them in their bid to innovate. It didn't make sense. If Best Practices inherently hold high burdens of knowledge, it is a losing battle.

The better solution is to find the first of these Best Practices, the first fruit, so to speak. The fruits that are ripe for the picking, low-hanging ideas that are already in existence yet remain hidden in virgin territory.

I call this Next Practice.

Next Practice is the first Best Practice. It has the advantage of starting from the point of a low burden of knowledge, where the chances of creating something new are high. Next Practice is often untested, experimental, and perhaps even counter-intuitive.

When we start from a low burden of knowledge, it takes much less effort to reach a breakthrough idea. It is analogous to climbing a short flight of stairs versus a long one. The former brings us to our goal a lot quicker because it requires much fewer resources while allowing us access to a plethora of low-hanging easy ideas. Climbing the stairway of a high burden of knowledge taxes us in multiple ways, yet leads us to scarce and more difficult ideas.

Best Practice
(Few Low Hanging Ideas)

Next Practice
(Many Low Hanging Ideas)

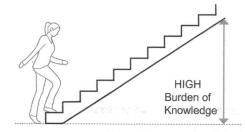

By definition, Next Practice is the *creation of an avant-garde course of action taken to solve critical challenges and problems faced by an organisation*. It is a solution that is generally accepted as surpassing any Best Practice currently in use because it produces results that are far superior.

Next Practice is in essence an *a priori* way of thinking, a Latin phrase used to mean knowledge or reasoning that starts from the point of predicting, an unproven deduction rather than a proven deduction. It is deductive logic. It goes from an idea, to an observation about the effects of that idea, and ends with a conclusion. Uncertainties are assumed as givens, with a constant fine-tuning of resource allocation to deal with these unknowns, so as to determine what can work better in future.

This is the direct opposite of Best Practice's *a posteriori* way of thinking, which uses past events and known facts to establish their reasoning. It is inductive logic. It has a preference for definitives, clear boundaries and looking to what has worked in the past, from observation to an idea that can explain what has been seen. Yet because hindsight is always 20/20, using Best Practice fails to capture the could-have-beens. As a result, businesses lose out on opportunity after opportunity.

Furthermore, unlike Best Practice, which has a static strategy of using a framework to determine the challenges to tackle, Next Practice reverses that with a dynamic strategy that defines a framework based on the challenges at hand. This flies against common use of innovation frameworks in business today to craft new products and services.

We have got the order of things wrong.

Best Practice	Next Practice
Managing the complicated	Embracing the complex
Closed system – deterministic relationship; certainties are the norm	Open system – probabilistic relationship; uncertainties are the norm
Past-oriented	Future-oriented
Exemplars exist for course of action to take	No exemplars exist for course of action to take
Static strategy – framework defines challenges	Dynamic strategy – challenges define framework
Inductive logic (from observation to idea) – *a posteriori* thinking	Deductive logic (from idea to observation) – *a priori* thinking
Amplifying strong signals	Amplifying weak signals

Next Practice methodology allows innovation to level up. Where Best Practice often only allows for incremental innovation, Next Practice leads to more types of innovation such as architectural, disruptive and radical innovation, because starting with new ideas is almost always richer than starting from the point of mere observations of old has-beens.

Let us take a deeper dive into the successes of Gojek and Microsoft, two giants that exploited areas of low burdens of knowledge and created their own Next Practices.

Case Study I: The Disruptive Innovator – Leaping Off the Incumbent

Gojek. The darling decacorn of Indonesia, born and bred in the fume-choked, soot-filled streets of Jakarta, where legendary traffic jams last for hours and commuters wait their lives away.

Nadiem Makarim started Gojek in 2010 when he saw a gaping void in the market caused by Indonesia's poor traffic infrastructure. Makarim, who was working for McKinsey at that point, had his personal car and chauffeur but relied often on ojeks, or motorcycle taxis, to get around Jakarta's infamous gridlock. Ojeks were cheap and could be found in every corner of the city, but operated in a highly inefficient manner. On a typical day, ojek riders could spend eight to ten hours waiting before landing an average of just six passengers. They were the lucky ones. The not so lucky riders could wait the entire day, with not a single passenger in sight.

Recalled Makarim: "So you've got this huge ineffi-ciency. Whenever you want an ojek, there will be no ojek. Whenever you don't want an ojek, they will be crowding pedestrian walkways all over Jakarta… they are one of the few service providers that have no influence over their abil-ity to get a customer. This was a market that was ripe for disruption."

In a bid to solve this glaring inefficiency and help remove some of the stresses facing ojek riders, Makarim decided to set up a small call centre to centralise and coordinate trans-actions between ojek riders and passengers. It received luke-warm response and did not take off as he had imagined. The tide turned in 2015 with funding by a private equity inves-tor. Gojek moved onto a digital platform, with the use of a mobile phone app that connected riders directly to passen-gers. That marked the start of a relentless march by Gojek to grow into the powerhouse it is today.

Indonesia at that point was already rapidly transform-ing into a digital nation, bypassing the desktop and laptop era to jump straight into the use of mobile phones for Inter-net connection. Riders and passengers easily transacted with each other because the system eradicated the need for them to haggle over prices, routes and timing.

What began as a social enterprise has now created mas-sive job opportunities for Indonesia's informal economy on a macro level.

"In most developing countries… I don't think unemployment is the core problem. It's under-employment. Under-employment is the inability to actually make as much money as you want to, to earn as much as you want, and this is what Gojek solves in a very, very elegant way because with a cheap Android handset, a helmet and motorcycle, people can work whenever they want, 24 hours a day, they can work from wherever they want," asserts Makarim. Gojek has, in essence, provided an opportunity for hundreds and thousands of Indonesians to become micro-entrepreneurs.

Gojek has since grown into a multi-service platform and digital payment group that serves over 170 million users across Southeast Asia. Currently valued at $10 billion, it announced its merger with Tokopedia in May 2021 to go public, in a move that will see a combined worth of $18 billion and create one of Southeast Asia's biggest tech conglomerates. Today, it already collaborates with over two million drivers and 900,000 small and medium-sized enterprises.

"We're coming to this point, a new technological paradigm that is the antithesis of this notion of scarcity. Gojek views the informal sector as an elastic resource that can expand in ways that we cannot even imagine. We hope to… show the world that social impact and scalable tech businesses are not only compatible but are, in fact, inseparable."

So how did a small call centre for just 20 drivers grow into the mammoth it is today in barely over a decade?

"We love to be either counter-intuitive or slightly counter to prevailing beliefs… we built a product around the frictions that an average person experiences in their day to day life. We are a platform that fixes things, we make things more efficient," explained Makarim, in a telling revelation of the secret behind innovation in today's economy.[10]

Gojek is more than counter-intuitive. It offered the Next Practice in solving Jakarta's traffic gridlock. Before Gojek came into the picture, the Best Practice to solve Jakarta's jams was to use taxis – a Blue Bird, to be precise.

The Blue Bird Group is the largest taxi company in Indonesia. Founded in 2001, it grew to take control of the taxi industry in the nation because of its own Next Practice when it offered an elegant solution to the transportation problems faced in chronically congested Indonesian cities. Want to be sure of a safe commute and fair rates? Hail a Blue Bird. The Blue Bird was king of the roads.[11]

But Gojek came along, and was able to do what no other company had done in usurping the incumbent transport company in Indonesia, not by trying to do better but by establishing the Next Practice with their motorcycle taxis. They created an entirely different market segment, with a focus on taking a completely different trajectory from where Blue Bird was heading, leapfrogging off the industry giant in order to help people better their lives.

Case Study 2: The Architectural Innovator – Unlearning, Relearning, and Looking Deeper[12]

Microsoft was once the juggernaut of the IT world. But something changed.

The undisputed technology titan and forerunner of the best and latest in IT, Microsoft had by the early 2000s lost much of its lustre and was in danger of becoming irrelevant. From its heyday of being the world's most valuable company in 1998, by 2013 the company was left straggling in the fast-growing fields of cloud, mobile and search. It had not caught on fast enough.

It was in danger of joining the ranks of big names that failed to innovate in time and had lost their way in the cut-throat world of business, such as Nokia, Kodak and BlackBerry.

Bill Gates accurately analysed what the future could be for Microsoft when he said in 1991: "This is a business where if you don't stay ahead, you can lose market share very quickly. We have to figure out what the opportunities for innovation will be, based on changes in hardware and changes in what users want, and then we have to implement those things very rapidly."

Fortunately, the narrative for Microsoft changed course. Like the proverbial phoenix rising from the ashes – with a timely change in leadership – Microsoft pulled itself back up to hit a record-breaking high in fiscal year 2020. Revenue went up from $86.6 billion in 2014 to $143 billion in

2020, while its stock price rose from $40 to over $200 in the same period. Under CEO Satya Nadella, the company made great strides in artificial intelligence and edge computing, while managing the largest commercial cloud business in the world.

How did Microsoft pull off such a monumental revival of its business, a rare feat that few other companies have achieved? The answer lay in Nadella's astute observation that he needed to transform the company from an outdated single software licence business for operating systems and productivity applications towards an SaaS (Software as a Service) business model. In 2014, Nadella announced the cloud-first strategy for the company to prevent another missed entry as had happened before in the mobile business.[13]

Nadella's intention to establish the Next Practice in Microsoft was made apparent during his Cloud First press briefing: "I think T.S. Eliot captured it best when he said that you should never cease from exploration, and at the end of all exploring you arrive where you started and know the place for the very first time. And for me that has been more true than ever before.

"And today marks that beginning of exploration for us... and everything that we do going forward is grounded in this worldview, which I describe as the world of ubiquitous computing and ambient intelligence. It's an amazing canvas for innovation and it's an amazing opportunity for growth for our company.

"When you think about the canvas itself, there are three aspects that really jump out. The first one is the world in the next five years and ten years is not going to be defined by the form factors that we know and love today but the variety of form factors that will come to be over the coming years."[14]

His metaphor of a canvas is a perfect one for Next Practice-building. A blank canvas allows for making marks according to whatever the imagination brings to mind. It was clear that back in 2014, the idea of ambient intelligence was not well known. So Nadella and his team stepped in to create their Next Practice.

Forming a senior leadership team (SLT) to spearhead a cultural renaissance within its 120,000-strong employee count, Nadella redefined Microsoft's mission: to empower every person and every organisation on the planet to do more and achieve more. This again was a Next Practice from Microsoft's previous mission as created by Bill Gates: a computer on every desk and in every home.

Said Nadella: "We need to be willing to lean into uncertainty, take risks and move quickly when we make mistakes, recognising failure happens along the way to mastery. And we need to be open to the ideas of others, where the success of others does not diminish our own."

The massive culture transformation freed Microsoft up to return to what it does best: being at the forefront of technology. More importantly, Microsoft was able to maintain focus not just on catching the first waves of technological advancements, but more precisely, on the forefront of technological advancement.

Like Gojek, Microsoft showed that it is possible for businesses, young or old, to discover the Next Practice by being and staying nimble, sometimes even creating a second or third Next Practice.

Finding Gardens of Eden

It is not easy to find the Next Practice.

Many businesses have tried, and failed, time and time again to innovate. Despite the best people, resources and ideas, they fall short of the mark. Why?

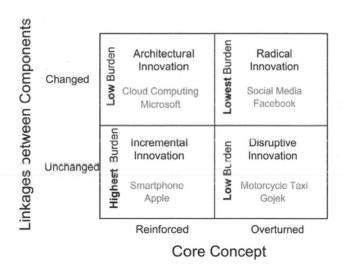

Types of Innovation
(modified from Henderson and Clark, 1990)

Looking at the figure above, my analysis is that these businesses remain at the Incremental Innovation quadrant. It is safe enough; the other three quadrants of Architectural Innovation, Disruptive Innovation and Radical Innovation appear as Danger Zones because of too much risk and unknowns.

What if they shift their perspectives to view these as Gardens of Eden instead? These are where new and young ideas are lush and plentiful, there is little to no competition, with less need for resources as a start point. It is a place to feast at, with so much for the taking.

It is possible, once we know the secret route.

I will show you the way.

CHAPTER TWO

Are We Born Innovative?

The digital world is a seductive one.

Facebook. Instagram. Twitter. TikTok. Massive Open Online Courses (MOOCs) offer free courses from the best institutions in the world, such as Harvard, Stanford, Berkeley, to anyone and everyone. Google is the de facto place we go when we want to search for just about any topic we can think of. YouTube is current king of how-to videos, having gained even more popularity during the Covid pandemic as more people turn to videos to replace face-to-face activities.

Together with a plethora of other social media platforms, websites and online games, these promise users endless hours of fun, entertainment, and yes, even learning. It is mind-boggling to think about how much we can access from anywhere in the world, with just a device and the Internet.

And spend hours, we certainly do, exploring this tantalising world.

More than 60% of the world's total population of 7.85 billion people use the Internet, with a majority of them – 92.8% to be exact – accessing it via their mobile devices. The average person spends 6 hours 56 minutes on the internet daily; that is almost a third of the day spent online. Of that, an average of 2 hours 30 minutes is spent on social media and messaging alone.[15]

With all that social interaction, albeit virtually, learning and playing online, it must be making people smarter, better informed and more connected. It must give us new inspiration, groundbreaking ideas, leading to more innovation and creativity, and possibly the discovery of more Next Practice.

Unfortunately not. Study after study show decreasing intelligence levels, and an increasing rate of digital addiction and mental health issues, all related to the overuse of digital technology.

Technology has brought about great strides for mankind. Without inventions in technology, we would never have computer systems that connect the world, nor cloud computing and artificial intelligence that have made our lives more productive and effective. Technology is here to stay, and it is not something we can live without. However, like a sharp knife, it needs to be handled properly. In the hands of one who knows how to master it, technology can only enhance and elevate what is done to even greater heights. But in the hands of an amateur, it can do great damage.

How can a seemingly good thing create chaos instead of helping us think better?

Before we answer this, we need to first understand how our brains are designed to think, and take a step back into the history of mankind's evolution.

Our Brain's Dual Systems

Nobel laureate Daniel Kahneman is a renowned economist and psychologist, particularly noted for his work in human rationality, judgement and decision-making. In his 2011 New York Times bestselling book and intellectual memoir, *Thinking, Fast and Slow*, Kahneman explored the assumptions behind what we understand about how our brains work, and introduced two characters: System 1 and System 2. These systems describe the two parts of the brain that think intuitively and deliberately, respectively.

Let us take a closer look at these two important characters that run everything from how we eat, walk and talk, to how we learn, think and make decisions.

System 1 is the lightning-fast agent of our brain. It requires little or no effort, operates automatically and quickly, and to an observer, is the intuitive, impulsive and freewheeling part that acts instinctively. It is the fun-loving partner you go with on spontaneous road trips, winging it with nary a plan or budget. Intuition and instincts are born from System 1.

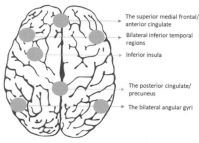

Areas of the Brain Affiliated with
System 1 Processing

The superior medial frontal/
anterior cingulate

Bilateral inferior temporal
regions

Inferior insula

The posterior cingulate/
precuneus

The bilateral angular gyri

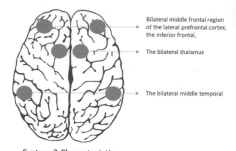

Areas of the Brain Affiliated with
System 2 Processing

Bilateral middle frontal region
of the lateral prefrontal cortex,
the inferior frontal,

The bilateral thalamus

The bilateral middle temporal

System 1 Characteristics

- Pattern Detection
- Automatic Operation
- Reduce/Simplification
- Emotion
- Multitasking
- Minimise Effort

System 2 Characteristics

- Pattern Recognition
- Controlled Operation
- Enlarge/Sophistication
- Logic
- Focusing
- Maximise Effort

System 1 vs System 2[16]

System 2, on the other hand, is the partner who pulls you back to earth and reminds you to do cost-benefit analyses of those road trips, weighs the pros and cons, and thinks through a million different scenarios. It is effortful and deliberate, and constructs thoughts in an orderly series of steps. Because it is slow, it can take in complex circumstances and maintain a certain level of concentration and exertion. Innovation and creativity are born from System 2.

Both systems are in operation when we are awake. System 1 generally runs the show, while System 2 stays on the sidelines in a low-effort mode. Driving to and from your child's ballet class? System 1 is in charge. Moving through your intense workout sequence? System 1 as well. Pulling your dog back onto the sidewalk to avoid that speeding cyclist? You guessed it. System 1 did the job of deciding. Automatic responses, snap decisions and daily tasks are all taken care of by System 1.

When System 1 runs into difficulty and does not have an answer for the paradigms and boundaries it is used to, System 2 gets mobilised. We suddenly become alert and focused, paying great attention to the issue at hand in order to find a solution. We go through each line item to balance the company's balance sheet; we painstakingly try to troubleshoot and fix the smart electronic home system that keeps rebooting for no apparent reason. And remember the time we felt like screaming at the driver who took the parking lot we were eyeing? System 2 was what kept us from losing self-control and getting into a fight.

When we think of how we operate, we think of ourselves as reasoning selves that make conscious choices and decisions, operating from System 1. It is surprising, therefore, to realise that System 2 is actually the main protagonist of the story.

This, however, poses a challenge. Because System 1 works automatically, errors of intuitive thought and biases are often difficult to prevent. Kahneman terms it the illusion of validity, a cognitive bias where people overestimate their ability to accurately interpret a situation – but end up making errors. The only way to override these errors is if System 2 constantly monitors System 1, which is impractical because it is much too slow and inefficient. So, System 1 prevails.

In essence, System 1 is the auto-pilot that keeps us running 95% of the time, while System 2 works intentionally and is activated 5% of the time. This means that when it comes to the daily division of labour, they operate like clockwork in a most efficient manner.

Minimal effort in thinking, optimal performance in operation.

Sounds good, yes? Not necessarily. Innovation requires going beyond the tried and tested boundaries of efficiency. If, as Kahneman puts it, our brains are designed for efficiency rather than effectiveness, implicit in this is the realisation that the human mind is already at a disadvantage when it comes to thinking deep and problem-solving.

Brain's Evolution to Conserve Energy, Act Fast

Over millions of years, the brain capacity of our human ancestors went through multiple transitions of growth, evolving to become what we are familiar with today.[17] These transitions led to the development of neural and cognitive advancements that have allowed *Homo sapiens* to act, talk and think intelligently, and with purpose.

This enhanced brain, however, is an energy-guzzler. Though weighing only approximately 1.4 kg, it uses a whopping 20% of the body's energy, with two-thirds of that going to firing neuron signals and the remaining one-third used for cell health maintenance.[18] That is a lot of energy for a rather small part of the body.

Our brain runs exclusively on the sugar glucose, with more strenuous cognitive activities needing more glucose than simpler activities. For example, if we are trying to work through the steps of solving a Rubik's Cube, the parts of our brain involved in problem-solving would use more energy than other areas of the brain would. Indeed, Harvard Medical School published a paper in 2016 that said, "The brain is dependent on sugar as its main fuel. It cannot be without it."[19]

To illustrate how much energy the brain burns, we can turn our attention to the world of chess and its grandmasters. In 2004, winner Rustam Kasimdzhanov walked away from the six-game world championship 17 pounds lighter. In October 2018, chess players monitored by Polar, a U.S.-based company that tracks heart rates, saw 21-year-old Russian grandmaster Mikhail Antipov burning 560 calories in two hours of sitting and playing chess. That was the equivalent of what tennis world champion Roger Federer would typically burn in an hour of singles tennis.

Yet, unlike muscles that can store excess carbohydrates, the brain is unable to squirrel away an energy reserve for when it needs it. It needs a constant supply of oxygen and energy to run properly, without which, neurons will begin shutting down quickly. While this seems to be a peculiarity, it actually helps the brain work better since stored energy cells would take up precious space in between neurons. Electrical signals would then need more energy to travel longer distances, thereby making the brain less efficient. So again, System 1 is the preferred mode to operate from since it takes up way less energy than System 2.

Furthermore, when we consider how System 1 evolved to become the dominant feature in our hierarchy of thinking, the 'fight or flight' nature of our limbic brain comes into focus. The amygdala in the limbic system is what forms a fast subconscious evaluation and response to keep us safe, bypassing the executive thinking brain completely.

In a nutshell, our brains are hardwired to work at an intuitive level and with minimal energy – both characteristics of System 1 – in order to ensure the survival of humankind.

> **"We think much less**
> **than we think we think..."**
> Daniel Kahneman

Environmental Destruction of Our Smartness

As if it is not enough that our brains prefer efficiency over effectiveness, the digital ecosystem around us works against boosting our ability to be creative and think smart.

It is in fact making us dumber.

The Flynn Effect points to a substantial and steady rise in IQ scores across the world in the 20th century (approximately three IQ points per decade) – up until 1975, when strangely they began to decline steeply (seven IQ points per decade).

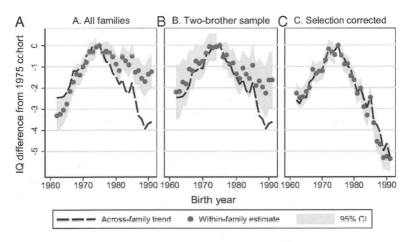

The Flynn Effect in IQ scores[20]

If we were to look at the history of computers and map it on the results of this study done on the Flynn Effect and its reversal, we will notice a rather curious, but unsurprising, trend. You guessed it. The drop in IQ scores coincided with the introduction of the workstation and personal computer.[21]

A 2017 study found that intelligence levels across Europe were falling, with Scandinavia and the UK as key places with declines observed since the mid-1990s. Researcher Michael Shayer said that since 1995, a "large social force has been interfering with children's development of thinking, getting larger each year". This force included the development of technology, such as game consoles and smartphones, which have altered the way that children communicate with each other.

"Take 14-year-olds in Britain. What 25% could do back in 1994, now only 5% can do," he added, citing math and science tests.[22]

What is an even more telling observation of our declining intelligence is data that shows Scholastic Assessment Test (SAT) scores are going south as well, in the areas of reading and writing.

SAT scores are used for college admissions; the brightest students from all over the world with the best SAT scores use them to gain entry to the top universities in the United States. These include students from the richest countries and families, with access to unlimited learning resources. Yet even then, scores are in decline. What could it possibly indicate then about the cognitive abilities of students over time?[23]

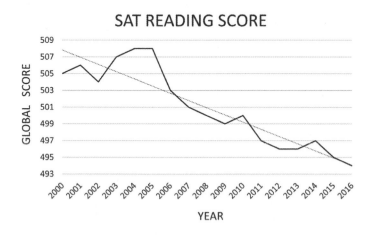

SAT READING SCORE

Algorithms Distort Our Thinking

"All of our minds can be hijacked. Our choices are not as free as we think they are."
Tristan Harris

Perhaps what is most chilling about the use of digital technology is how it is messing with our thinking and ability to focus.

Tristan Harris is a former Google employee and now whistle-blower with a reputation for being "the closest thing Silicon Valley has to a conscience". He is a strong advocate for awareness on the invisible ways in which Silicon Valley is shaping the thoughts of billions of people on the planet, people who have neither choice nor understanding about how technology companies wield such influence over them. "A handful of people, working at a handful of technology companies, through their choices will steer what a billion people are thinking today," he said at a 2017 TED talk in Vancouver.[24]

Unbeknownst to the man on the street, our digital usage is not under our control as much as we think it is. The second we turn on our devices, we are silently being led down paths determined by someone else.

Take YouTube for example.

There are 2.3 billion YouTube users globally, as of 2021. Once they start watching the first video, their playlist is automatically recommended, and played, according to an algorithm that will identify and rank 20 'Up Next' clips that are deemed relevant to the previous video. That algorithm is a secret formula, proprietary software that is designed to keep users watching, and is the single most important engine of YouTube's growth.

It harnesses what is known as association rule learning, which is a rule-based machine learning method for discovering interesting relations between variables in large databases. YouTube learns about the connections between interests, attention levels and users, and builds links.

Guillaume Chaslot, a French computer programmer with a PhD in artificial intelligence, used to be a YouTube software engineer working on this algorithm. In his time at Google, Chaslot realised that the priorities YouTube gives its algorithms are dangerously skewed to show the most controversial and edgy content.

The focus is on showing people videos they find irresistible; people end up viewing content that reinforces their existing view of the world, and going deeper into one area of knowledge without gaining perspective of alternative views. It ends up building a knowledge trap.

Said Chaslot, "YouTube is something that looks like reality, but it is distorted to make you spend more time online. The recommendation algorithm is not optimising for what is truthful, or balanced, or healthy for democracy." It basically distorts truth.

The use of algorithms and corralling users into taking a certain path of viewing and thinking are prevalent in practically all the big technology companies. Their mission is simple: to keep users on their product for as long as possible.

And we, as digital tech users, get sucked in further and further down the rabbit hole of curated content that was never ours to manage right from the start — straight into what is known as filter bubbles.

Eli Pariser, executive of Upworthy, activist and author, explains in his book *Filter Bubbles* how Google searches are highly dependent on who is making those searches. Two people can search for the same thing, but get completely different results. Why?

Pariser shares that the Internet feeds us with what it thinks we want based on all the data it collects about us, and creates a "personal ecosystem of information", insulating us from any sort of cognitive dissonance by limiting what we see. These algorithms curate "a unique universe of information for each of us... which fundamentally alters the way we encounter ideas and information."[25]

> **"What's in your filter bubble depends on who you are and it depends on what you do. But you don't decide what gets in. And more importantly, you don't actually see what gets edited out."**
>
> Eli Pariser[26]

From filter bubbles, we then find ourselves inside an echo chamber of hearing only what we agree with – and we end up distrusting everyone else on the outside. Echo chambers isolate us, not by cutting off our lines of communication to the world, but by changing whom we trust. Outside views are discredited, and we grow more and more resistant to views that differ from ours. These echo chambers can span a wide spectrum of segments in society, from parenting forums and vaccinations, to nutritional methods and even exercise regimes.[27]

As far back as 1996, MIT researchers Marshall Van Alstyne and Erik Brynjolfsson warned of a potential dark side to our digitally interconnected world: "Individuals empowered to screen out material that does not conform to their existing preferences may form virtual cliques, insulate themselves from opposing points of view, and reinforce their biases. Internet users can seek out interactions with like-minded individuals who have similar values, and thus become less likely to trust important decisions to people whose values differ from their own." That prediction has already come true.

Factual accuracy online can no longer be assumed as a given. Unlike traditional print media which are still subject to some form of editing for objectivity, online media and the ability for just about anyone to put up content online means a proliferation of information that tells the audience exactly what it wants to hear. With an estimated 61% of millennials getting their news from social media, this further triggers algorithms that curate their feeds to give them what they agree with, and keep away alternative points of view.[28]

Once inside these echo chambers, we end up thinking that everyone thinks like us, and forget other perspectives exist. Both filter bubbles and echo chambers trap us further in our System 1 way of thinking. That cannot be good at all.

Addictive Digital Use Reinforces System 1

You know the micro-buzz of pleasure you secretly feel whenever you get a notification that yet another comment has been made on your post? That was the result of a series of carefully calibrated steps and techniques used to get you hooked and build your digital usage habit.

Technology companies exploit our human desires for approval, affirmation and social acceptance through a variety of tools and features such as the Facebook 'Like' button, push notifications and the auto-play function on YouTube. But perhaps the most alluring techniques play on the psychological susceptibility that makes gambling so irresistible: variable rewards.

Introduced by psychologist B.F. Skinner in the 1930s, the variable reward schedule exposes users to reward-associated stimuli and dispenses rewards after a varying number of responses. No one can really predict when the reward will come. It is the possibility of reward vs disappointment that makes the action extremely compelling. Each time a response to a stimulus results in a reward, an association is made and the process strengthens the neural pathway between neurons, thereby increasing the intensity at which a response is made to the stimuli in future.

So whenever we swipe our phones and see an avalanche of 'Likes', or pull-and-refresh our screens to discover new posts on our feed, we are rewarded with a hit of dopamine, much like what cocaine addicts and gamblers get high on.

"Each time you're swiping down, it's like a slot machine," Harris says. "You don't know what's coming next." Platforms like Facebook, Snapchat, and Instagram leverage the very same neural circuitry as drugs and gambling to keep us using their products as much as possible.[29] Money and resources are pumped by technology companies into finding ways that will keep us engaged. It is a race for our attention and time, in order to chase the advertising dollar.

Yet even as billions of Internet users continue to swipe, refresh and scroll, many younger technologists who created all the little seductive features that keep users addicted are themselves weaning off their own products, and sending their children to elite Silicon Valley schools where iPhones, iPads and even laptops are banned. These tech executives are disconnecting from the Internet.

It is an irony that must not be lost on us.

There is a mounting and urgent wave of concern that users are not only getting addicted to technology, but are also being unwittingly trained to be in a state of "continuous partial attention", where their ability to focus is severely limited, and intelligence affected. A study[30] showed that the mere presence of smartphones damages cognitive capacity, even if people are able to resist the temptation to touch their phones, and even if the devices are switched off. "Everyone is distracted," said tech executive Justin Rosenstein. "All of the time."[31]

Our usage of digital technology has fundamentally made us prisoners, trapped to think and behave from a place of automaticity. While a System 1 way of functioning works for the day-to-day, it is the antithesis of deep thinking and problem-solving. We cannot begin to see differently if we remain stuck.

Founder and CEO of Telegram, Pavel Durov, has also sharply criticised social media recommendation algorithms. Said Durov on his Telegram channel: "To be creative and productive, we must first clear from our minds the sticky mud of irrelevant content with which 'recommendation algorithms' flood it on a daily basis. If we are to reclaim our creative freedom, we must first take back control of our minds."[32]

We know our minds are powerful machines capable of generating brilliance. But for this to happen, we must nourish our brains with high-quality information.

"It is unfortunate that most people prefer to feed their minds not with real-life facts that can let us change the world, but with random Netflix series or TikTok videos. On a deep level, our brain can't tell fiction from reality, so the abundance of digital entertainment keeps our subconscious mind busy producing solutions to problems that do not exist."[33]

Furthermore, what we think we know may not be what we truly know. Our consciousness is like a 'user illusion', according to the Eliminativist view. Say in the case of optical illusions, our brain fills in what is not there, sees movement where there isn't any, and basically makes things up. This is very much in line with Kahneman's illusion of understanding. If both our conscious and subconscious minds are stuck in opacity, how will we ever be able to think creatively?[34]

Is there a way out of this mess? Can we escape the prison of System 1 and access System 2 more?

CHAPTER THREE

The Effulgent Mind

"This man is a mathematical genius."

Written by Carnegie Mellon professor Richard Duffin in 1948, this sentence was part of a short and succinct recommendation letter to the chairman of the mathematics department at Princeton for John Forbes Nash Jr's application to graduate school.[35]

Six words that can barely contain the sheer brilliance of one man.

John Nash, considered to be one of the brightest and sharpest mathematicians of the 21st century, illuminated our understanding of the fundamental ways people make decisions and take chances in daily life, and made groundbreaking contributions to the advancement of knowledge in the field of economics with his theories. He was known for originality in thought and fearlessness in tackling acutely difficult problems few others dared attempt. His work in game theory, eventually known as the Nash Equilibrium, won him the Nobel Prize in 1994.[36]

The Nash Equilibrium is now prevalent in virtually any situation in life where there is a need to make strategic choices in order to obtain a desired outcome – be it in investments and business, sports, politics or even playing board games. We probably all use the Nash Equilibrium on a daily basis without even realising it. The last time we chose paper in a game of rock-paper-scissors because we thought our opponent would choose rock was really the Nash Equilibrium at play!

Nash's life and work eventually inspired Sylvia Nasar's biography, *A Beautiful Mind*, and a 2001 film of the same name that went on to gross over $313 million worldwide and win four Academy Awards.

Not only was his work in game theory a game changer, Nash also had an almost incomprehensible ability to crack enemy codes and establish new robust ones during his stint with the National Security Agency of the United States. It was as if his brain worked on a different spectrum, dimension and level from the average person.

And perhaps it was indeed accessing a different part of the brain and way of thinking from the way we usually think.

A different sort of brain almost. A radiance.

An effulgent brain.

Brilliance of the Effulgent Brain

Kahneman explored the idea of System 2 as the boss of the brain. While System 1 may be the more visible star of the show in daily life, the role of 'CEO of thought' is actually played by System 2.

System 1 may propose, but it is System 2 that disposes. Its deliberate and rational nature, albeit coupled with an inherently lazy nature that prefers to stay under the radar when it can, is what allows it to make effective meaningful decisions.

This seamless dance between both systems is well-illustrated by how the fluid movements between them help in medical decision-making. The job of a medical professional relies greatly on the ability to rapidly and accurately diagnose, treat and improvise during stressful situations. However, medical errors and mistakes happen when these professionals operate from a place of heuristic shortcuts and intrinsic biases.

When medical students are new to the wards, they rely greatly on System 2 because there is a need to purposefully get familiar with protocols, obtain patient history, conduct detailed physical examinations and the like. Once these become second nature, System 1 takes over the day-to-day operations, leaving System 2 to move on towards assessment of plans for the management of acutely ill patients and other medical care goals.

In summary, the cumulative effect of medical training allows for medical professionals to consolidate their daily tasks from an intentional System 2 to an automatic System 1 way of thinking, in order for System 2 to reclaim its role as a thinking CEO. It is a symbiotic relationship; one cannot exist without the other, but with System 2 as ultimately the one in charge.[37]

Let us go on an exploration of the features of this beautiful luminous mind.

Feature 1: Paying Attention

**"The quieter you become,
the more you are able to hear."**
Rumi, 13th-century Persian poet

Enter any Buddhist temple in Japan, and gradually, a peace descends.

You walk the grounds and notice the almost-requisite minimalist Zen garden that evokes a spirit of tranquillity. The architecture of the buildings draws your attention to another world beyond the now, and invites a pause, a reflection. There is a quiet.

Your senses come alive one by one, and you find yourself becoming aware. Of the curves and the wood, the stones and the water. You start to notice the grandest structure and the littlest detail carved into that structure, hear the crunch of gravel beneath your feet, feel the rough-hewn wood of the bridge as you cross the stream in the garden.

You start to pay attention.

When we pay attention, it is our window to the world. In Kahneman's scheme of Systems 1 and 2, the latter helps us understand the world by deliberately and consciously paying attention to input from the world and making sense of it. It carefully thinks.

However, this kind of effort does not come naturally, as we have already seen of System 2's preference for lying low unless called upon. It requires continuous exertion and effort. We must direct our attention to the task and situation at hand. If we are either not ready to be attentive, or our attention is on the wrong thing, performance suffers.

Without System 2, such attention simply cannot take place. System 1 is incapable of handling this role because it operates from a point of automatic response. Without System 2 to enable attention, we effectively become blind to what is happening around us, falling back instead on what we think we already know.

In the book *The Invisible Gorilla*, Professors Christopher Chabris and Daniel Simons document one of the most well-known experiments in the field of psychology that demonstrates our minds do not work the way we think they do. The experiment shows six people – three in white shirts and three in black shirts – passing basketballs around. The viewer is asked to keep a silent count of the number of passes made by the people in white shirts. Easy enough task, isn't it? Here's the twist: at some point, a person in a gorilla suit strolls into the middle of the group, looks at the camera and thumps its chest, then unhurriedly walks out. It spends a total of nine seconds on the screen. Yet out of all the respondents who watched the video, a whopping half of them did not see the gorilla at all. Yes, the human-sized gorilla that did a little jig in plain sight, before exiting. How is that even possible?

Without paying attention, we are effectively blind, even to the most obvious stimuli. We think we experience the world the way it is, but how we respond to it often stems from sometimes-faulty instinct and intuition. Otherwise known as System 1. We end up being deceived into thinking we are logical and rational in our perception of the world and subsequent decisions made, only to realise that we have been waylaid by our System 1 way of thinking.

By paying attention, we will activate our System 2 more often, so we are no longer blind to our own blindness. We come alive, see more, take in more, become more intentional in the way we interact with the world.

Feature 2: Reflection

Self-made billionaire Richard Branson, founder of the Virgin Group, is a strong believer in taking time off from work and going on vacations.

"When you go on vacation, your routine is interrupted; the places you go and the new people you meet can inspire you in unexpected ways," said Branson. "I make sure that I disconnect by leaving my smartphone at home or in the hotel room for as long as possible – days, if I can – and bringing a notepad and pen with me instead.

"Freed from the daily stresses of my working life, I find that I am more likely to have new insights into old problems and other flashes of inspiration."[38]

Taking time away from the grind to purposefully cultivate the art of reflection is a habit that a growing number of CEOs and leaders are building in order to unlearn, relearn and improve.

Some meditate, like the late Lee Kuan Yew, founding father of modern Singapore, who meditated daily under the tutelage of Catholic Benedictine monk Fr. Laurence Freeman and credited meditation with helping him gain clarity and think through options when he had to make tough decisions. Others like Warren Buffett, Bill Gates and Mark Zuckerberg are voracious readers of different genres to gain new and wider perspectives. Still others disconnect for scheduled thinking time on questions that prompt reflection.

What is reflection?

It is when we consciously consider and analyse our beliefs and actions for the purpose of learning. This is System 2 at work. When we reflect, our brains are able to pause in the middle of chaos, untangle knots and sort through different observations and experiences, make connections between seemingly unrelated pieces of information and create meaning. This meaning then becomes learning, and acts as a basis for future actions taken by System 1.[39]

Instead of acting from a position of reflex – one that is guided by impulsivity, emotion, automaticity and perhaps even reckless simplification – reflection helps us move into a space where we are focused and in control. Choices are made with reason and logic, guided by a sophisticated prudence that weighs options.[40]

We can certainly think of many situations in our lives where the ability to reflect before jumping into something is highly advantageous, be it deciding on what career path to take or which person to marry.

Reflective thinking is also synonymous with critical thinking.

Researcher Jonathan Haber in his book *Critical Thinking* defines it as structured thinking with three objectives:

1. Making clear what we are thinking
2. Making transparent the reasons behind what we believe
3. Having the ability to determine if our beliefs are justified

Often cited as an essential 21st-century skill, critical thinking is lauded as one of the keys to success in school and work. Consider this: without it, societies would break down into total chaos in the face of fake news, incorrect conclusions and decisions based on emotion rather than reason. Nations would square off against each other in battle. All of this if you and I are incapable of thinking critically and logically.

Without critical thinking, analytical thinking will also be hard to manage. We will find it hard to dissect and study a problem in a logical manner to find the answer or solution. There will be little clarity in what we are even thinking.[41]

Scientifically, it is proven that analytical thinking exists, due to greater frontal theta activity in the brain and reduced parietal alpha activity, which suggests the recruitment of cognitive control, working memory, and focused attention. A child working on a mathematical problem would display such brain activity (System 2 slow deliberate thinking) versus an adult who is driving a car with little attention to the mechanics of it (System 1 fast automatic thinking).[42]

In a nutshell, the ability of the effulgent mind to reflect allows us to think critically, and finally analytically.

Feature 3: Slowing Down

Nestled among the lush greenery by the Ayung River, Bali, sits a grove of bamboo structures that house the Green School Bali.

As a private international school for pre-schoolers all the way to high-schoolers, Green School offers a liberating learning environment that is holistic, community-integrated and one that is disrupting traditional notions of what education should be.

Said Glenn Chickering, Head of Faculty: "Traditional schooling is often guilty of rushing their curriculum and pressuring students to know it all and not wait for other learners to catch up. But for me, content is a great vehicle for developing skills, and to do that, we need to slow down, make mistakes and reflect on them."[43]

Perhaps it is a sign of the times, but globally, the movement towards slowing down has been gaining traction, with a shift in speed towards food, parenting, fashion, technology, among a plethora of other areas. People are increasingly realising – and relishing – the power of slow and its benefits.

For slowness brings with it so many good things and lessons to learn.

Like the ability to manage cognitive reappraisal.

A group of MIT neuroscientists found in 2014 that the human mind requires just 13 milliseconds to process information from an image. A blink of an eye takes 100 milliseconds. The flap of a hummingbird's wing takes about 80 milliseconds. Which makes 13 milliseconds for us to react really, really fast.

Therefore cognitive reappraisal is extremely helpful to stop us in our tracks. This brain process gives us the ability to pause and re-evaluate our initial reaction to a situation. It is what allows us to not assume our spouse is being unhelpful and then getting upset, but instead regulate our emotions and assessment to respond more appropriately, so we realise later that he or she just did not hear our request to wash the dishes.

Neuroimaging studies have consistently shown that correlations in activity between brain regions evolve over time. Spatial patterns are formed, dissolved and reformed. The synchrony of brain activity changes over an ultra-slow timescale. Even the brain at rest works to form and dissolve multiple communities of harmonised brain regions. Our brains need slowness in order to function better, which is very much in line with Kahneman's systems of thinking and the optimal operating environment.[44]

System 2 needs to slow down in order to deliberate and reason. That longer time to deliberate allows for a surer outcome, versus a faster and looser way of thinking that is as inattentive as it is intuitive. System 2 with its ability to think critically requires time to work.

The common fallacy is that the quicker things change, the quicker we need to think. We end up spending so much time trying to solve problems, re-engineer, meet deadlines and the like that there is no time left to think deeply and slowly.

President of Kenyon College, Sean Decatur, is one such proponent of slow thinking. In a 2020 LinkedIn article on the importance of slow thinking, he noted that while fast thinking is needed for survival, the VUCA (volatile, uncertain, complex and ambiguous) nature of the Covid-19 pandemic has revealed an inability to cope effectively with crisis simply by thinking fast.

"Circumstances change quickly and unexpectedly; the

future is clouded by uncertainty; solutions to problems are often ambiguous, and involve integration of complex, multi-disciplinary concepts. Fast-thinking breaks down under these circumstances. When we have not experienced life in a pandemic, we cannot rely on our fast-thinking – actions based on heuristic analysis of cues around us – to guide our behavior or our survival."

Rather, slow thinking is the key to survival.

"Fast-thinking may help make us human, but slow-thinking completes our humanity. Hearing is fast – even hearing to extract information is fast – but listening is slow. Reaction with emotion is fast, but reaction with empathy – understanding of the experiences of others – is slow."

The crucial point is this: slow thinking isn't just about learning.

"Slow-thinking often begins with unlearning," reflects Decatur. "Looking at the world in a different way – taking in new perspectives, working to resolve contradictory information, just taking time to process our thoughts and take measure of the act of thinking itself – all of these allow us to reshape the brain physically, to dismantle neurological connections and build new ones. Learning alone is not enough; we must complement learning with unlearning and relearning."

This kind of unlearning and relearning is helped in a big way by slowing down in both looking and knowing.

Psychologist Guy Claxton observed in his book *Hare*

Brain, Tortoise Mind that in the midst of chaos and complex situations, slow knowing is even more important. "One needs to be able to soak up experience of complex domains – such as human relationships – through one's pores, and to extract subtle, contingent patters that are latent within it," said Claxton. "And to do that one needs to be able to attend to a whole range of situations patiently without comprehension; to resist the temptation to foreclose on what that experience may have to teach."

Claxton references poet John Keats' mention of negative capability, which is the ability to wait attentively, even in a situation of incomprehension: "To wait in this kind of way requires a kind of inner security; the confidence that one may lose clarity and control without losing one's self."[45]

A good example of where slowing down the need to know is in the classroom. With eight seconds as the latest estimated length of the human attention span, teachers are constantly sandwiched between the need for speed and the need for understanding. Increasingly though, guiding students towards acquiring 21st-century skills such as critical thinking and creativity means taking time in the classroom. Students need to be given the chance to look slowly and practise observing details over time, so that they are able to move beyond a first impression and create a more immersive experience with a text, an idea, a piece of art, or any other kind of object. It's a practice that clears a space for students to hold and appreciate the richness of the world we live in.[46]

Ultimately, slowing down gives us clarity. It gives us a complete picture, because we no longer miss out on crucial details in our haste.

Feature 4: Making Decisions

The first case of Covid-19 was reported in Wuhan, China, on 31 December 2019.

Within the next few weeks and months, what seemed like just a new virus soon spread uncontrollably through practically every city and country, bringing the world to its knees. The virus was unstoppable, and no one seemed to know what to do to rein it in.

It was a global crisis that threatened to destroy mankind.

For world leaders and ordinary citizens alike, making good decisions in this unfamiliar terrain that changed by the day was tough. How do we pivot a business hit by Covid-19? Do we vaccinate ourselves with the new mRNA vaccines or wait it out? How do we support workers and students who have their work and learning totally disrupted?

Enter System 2 again, the hero of the day.

In the face of panic and chaos, the worst thing to do was to make snap decisions in a bid to avoid the deadly threat of Covid-19. Making thoughtful decisions so that System 2 could take over in thinking and making meaningful assessments was really what helped save us from defeat by the virus.

We know that the intuitive System 1 will, when faced with a problem, do the best it can by accessing relevant experience and finding an appropriate solution for it. However, when the problem is a difficult one in which System 1 has no previous experience solving, the brain automatically substitutes the difficult question with an easier and related one to answer. If the question is, "Should I pivot my business online to adapt to the new Covid-19 norms?", the easier question to answer is, "Do I like doing business online?"

Said Kahneman about this propensity for substitution: "This is the essence of intuitive heuristics: when faced with a difficult question, we often answer an easier one instead, usually without noticing the substitution."

An example of the enormous consequences of making serious misjudgements was in the case of the 1986 Space Shuttle *Challenger* disaster. The media, the country, the world were all waiting for the launch and that put tremendous pressure on NASA to go ahead, despite evidence of problems with the ship's O-rings in cold weather. The Principle of Bounded Rationality states that we only have that much time and energy to make a prudent decision; in this case, NASA, under scrutiny and time constraints, was bounded from making a rational decision. Working under System 1, it went ahead with a poor decision to launch, and it ended in a real tragedy.[47] Engaging System 2 thinking could have prevented this disaster from taking place.

Thus, if we insist on engaging our System 2 to tackle difficult questions head-on, and avoid the temptation of fast thinking, we allow System 2 to rise to the fore, resulting in better decisions rather than reactive responses simply because of an Instagram post, newspaper headline or something someone said.[48]

Look at any leader's task of making decisions. It is the toughest and riskiest because so much lies at stake. A bad decision can derail an entire career, a company's carefully laid-out plans, years of goodwill and reputation. What runs in a leader's mind, be it psychological traps or heuristic flaws, they are System 1's invisible arms and can determine the success or failure of that decision. We need System 2.[49]

Wise decisions, astute decisions, or comprehensive decisions may never exist without System 2.

Feature 5: Casting Doubt

The devil's advocate, or *advocatus diaboli* in Latin, was once an official position in the Catholic Church. As a promoter of the faith whose job was to argue against the canonisation of a candidate and look for holes in the evidence, it was a much-needed role to ensure the candidate was worthy of sainthood.

In the same way, each of us too has a devil's advocate we can rely on to keep us on our toes when we need to make assessments and sound decisions. This is highly necessary because our brains naturally like to take shortcuts. We are cognitive misers and our System 1 finds it too much mental work to make a thorough evaluation every time we need to make a decision. It then results in a narrow thinking that hems in our perspectives, weaves a compelling story, and gives us great confidence in making a decision.[50]

So we end up doing things like selectively picking information that confirms our existing beliefs and ideas, rather than question them or look for new ones. This cherry-picking is known as confirmation bias, and is why you and your friend can have opposing views on which political party is the best, see the same evidence, and still feel each one of you is right based on the evidence. Cognitive bias is most at play in the case of ingrained, ideological or emotionally charged views. When we fail to assess information in an unbiased way, it can lead to serious misjudgements.[51]

Science is a subject usually seen as a process that uncovers facts and certainties. Yet the paradox is that science doubts. It challenges hypotheses and facts in order to yield new knowledge and questions; it changes our understanding of the world whenever new findings show that our prior understanding was wrong. Doubt creates new paradigms. That is the true driver behind the power of science. Doubt overturns the old to make way for the new.[52]

System 2 opens our mind to a whole new world and multiple perspectives because of its ability to question and doubt. This power of doubt invites us to move beyond jumping to conclusions. As Kahneman puts it, we go from believing that "what you see is all there is" to *what you see with deliberate looking is a whole lot more,* and therefore think harder and deeper.

Wharton School professor Adam Grant in his book, *Think Again: The Power of Knowing What You Don't Know*, invites us to explore how rethinking happens; to pivot away from System 1 by letting go of knowledge and opinions that are no longer serving us well, towards a flexible System 2 that considers possibilities rather than certainties.

★   ★   ★

In 2016, tens of millions across the world played spectator to a historic match between Lee Sedol, recognised as one of the world's best in the game Go, and AlphaGo, an AI machine designed by researchers at DeepMind, a Google-owned London AI lab.

The match saw AlphaGo pulling a stunning Move 37 in Game 2 that crushingly defeated Sedol and shocked everyone; never before had such a beautiful move been made. While the machine went on to win Game 3 as well and the entire match in the best-of-five format, Sedol turned the tables around in the very next Game 4 to beat AlphaGo in Move 78, which has been dubbed "God's touch". It took AlphaGo by surprise. That one-in-ten-thousand-chance move exactly mirrored AlphaGo's one-in-ten-thousand-chance move.

While AlphaGo had years of training to study the human method of play, Sedol had only that one match to learn from the machine, and do exactly what AlphaGo did.

It proved that the genius of the effulgent human mind can never be overcome by a machine.[53]

Human beings as a species are increasingly having the singularity of our intelligence challenged. Computers and robots have been overtaking us in numerous ways – be it in playing chess or predicting social media algorithms. It may seem dismal, if not for one thing. They don't know what they don't know.

Us? We have the ability to use metacognition strategies. To think about our thinking, see what we don't know, to make meaning, and finally learn, unlearn and relearn. And that is the beauty of our effulgent mind, the same kind of mind that John Forbes Nash Jr inspired the world with.[54]

Creating Next Practice from the Inside Out

Brains Can Get Smarter

> **"Intellectual growth should commence at birth and cease only at death."**
> Albert Einstein

Imagine you are visiting a new town and your mobile phone is not working. You cannot rely on the GPS so you fish out a physical copy of the road map.

Chances are, it would feel foreign and frankly, rather tricky trying to figure out the route on your own, without having a voice give you specific instructions on when to next make a left turn. The lack of practice with the physical map and the resulting difficulty in using it suggest that the neural pathways associated with map-reading have weakened or even disappeared over time.

After struggling for a while with map orientation, reading road signs and matching map to road, the process suddenly seems somewhat easier and familiar. Voila! You are once again navigating the roads like a pro.

Welcome to the world of neuroplasticity.

Neuroplasticity is the ability of the brain to create, strengthen, weaken or dismantle neural connections. New neural connections can form, old connections can modify or rewire, based on neuronal activity and experiences. Neural pathways grow and get reorganised, in a fascinating dance of neurons, axons and synapses.

It used to be common understanding that brains stop growing once a child grows up. Neuroscience studies have since shown that our brains can continue to grow and make new connections throughout life. This is good news!

If we want to become smarter, at any age, we can. What we need are the right exercises to maintain our brains' plasticity.

There are two types of neuroplasticity – functional and structural. The former has often been cast into the spotlight because of its role in healing from brain damage, e.g. from an accident. Brain functions move from the damaged parts to the non-damaged parts so that the person can learn how to use those functions again.

Structural neuroplasticity, however, is what I want to examine in greater detail, because it features greatly in our ability to find the Next Practice.

This type of plasticity is built by memories and experiences. We see it at work when we look at London taxi drivers. In order to pass the test to be a taxi driver in London, applicants undergo 3-4 years of training and need to pass a gruelling series of exams that will test their memory on a labyrinth of 25,000 streets within a 10 km radius of Charing Cross train station, as well as thousands of tourist attractions and hot spots.[55]

The result of that? Ballooned hippocampuses in the brains of these taxi drivers, which were either due to new neurons grown, or new neural connections made between neurons. These taxi drivers essentially became smarter by growing their brains.

In a 2019 joint study between Stanford and Arizona State University, it was observed that the hippocampus was responsible for encoding associations between relevant features of the environment, with these associations then used for learning and forming new memories. Conventional convergent and divergent thinking lack this ability to make associations between two or more seemingly unrelated ideas, then recombining them into a plausible idea.[56]

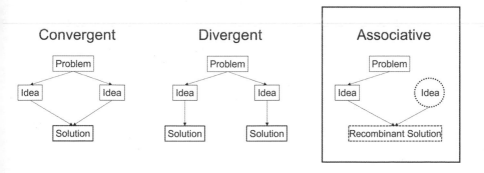

So, there is a way to become smarter and more innovative, to break out of the trap of System 1.

There is just one last step. Neuroplasticity requires strengthening something in particular to achieve its full potential.

Brain Lessons 101

Let us take a peek into the brain of our subject, Susan.

Susan Robert, 39, is a data analyst who spends her weekends meandering through the busy streets and back alleys of different neighbourhoods to take photos. She has created quite a stir in the art scene for her experimental marriage of avant-garde photography and glass sculpturing.

Looking at Susan's brain when she is working with photography and glass, we will see that countless neurons are firing off electrical signals and forming axonal connections among themselves. These connections, known as neural pathways, join one neuron to another to transmit information.

As a self-taught photographer and artist, Susan had to learn photography techniques from scratch, immerse herself in glass sculpting sessions, and figure out how to put both together to produce beautiful art sculptures. She may not have known it, but all that learning of new things and pushing past her comfort zone of knowledge was in effect making her smarter.

Building Neural Pathways and Myelination

Our brains comprise approximately 85 billion neurons. That is a lot of nerve cells packed into a relatively small area.

Each neuron is like a messenger that sends information, via nerve impulses, to other neurons. These signals travel through the nerves – since the neurons are part of the central nervous system – through the body to allow you to do everything you do, whether it is talking, eating or trying to thread that needle for the umpteenth time. Each neuron can connect with up to 10,000 other neurons in your brain, forming what looks like a dense spider web.[57]

These connections between neurons are known as neural pathways, or synapses. Psychologist Donald Hebb famously coined the phrase "neurons that fire together wire together" to explain these pathways. When we learn something new and repeat it, the various neurons involved in that process repeatedly fire together in a new sequence, and wire together as a collective.

The repeated firing signals that the new neural pathways are important. So each time we practise something intentionally and deeply, each time we struggle to overcome a challenge or learn a new technique, the neural pathways are forged and strengthened. Deep structural changes are made because of our learning and experiences. We are in effect reshaping our brains.[58]

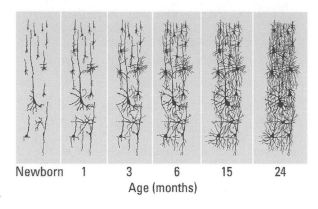

Newborn 1   3   6   15   24
Age (months)

Synaptic density over time[59]

Our brain transformation does not stop there. The neurons that keep firing to build neural pathways also send signals to do something else.

Myelination.

Dissect a brain and you will find what looks like white fibre bundles. These are actually neural pathways that are coated with a fatty substance called myelin, which act as electrical insulation. Much like the protective insulation on electrical wires in our homes, myelin keeps nerve impulses from leaking out of the synapses.

Imagine the thick insulation around aircon cables. They protect the copper wire and transmission of cold air within. Neither heat nor any other disruptions will result in a loss of temperature as the cold air is delivered through the blower into your room. Same for the brain.

Myelin also boosts the transmission speed at which

messages travel through the brain. Don't we all want souped-up brains? The higher quality the myelin, the faster the messages travel – heavily myelinated pathways are up to 300 times faster. It is the equivalent of travelling on horseback versus jet plane. These pathways are optimised for speed and efficiency, and end up as the default behaviour since the brain always chooses the most myelinated pathways.[60]

When Albert Einstein died in 1955, his brain was preserved, but due to the technology at that point, they did not discover anything of particular interest. However, 25 years later, in 1980, with better technology and knowledge of the brain, scientists examined his brain again to look for the secret to being a genius. Instead of finding extremely high numbers of neurons, they found a greater than normal amount of white fatty substances. Einstein's brain had myelin in abundance. It was further packed with more glial cells that could not only produce more myelin, but also supply neurons with nutrient energy.[61]

Something else that set his brain apart from normal human brains was how the different lobes and two hemispheres were more interconnected. The right frontal lobe, which controls higher thinking, had more folds than normal. The corpus callosum was thicker than usual, which meant more and better communication and connection between different parts of his brain.

It probably meant that when Einstein had an idea, it bypassed a strict logical sequence and transmitted through many areas of his brain, be it mathematical, spatial, verbal or visual. Chances are, he daydreamed often to allow for such fluid, seemingly random, transmissions to take place.

In a nutshell, myelin is what we want in order to get smarter, think faster, work better. To reach genius level.

And this magic of myelination happens when the constantly firing neurons alert a group of brain cells called oligodendrocytes to extend tentacle-like arms towards parts of the neural pathways, grab them, and begin to concentrically wrap tongues of membrane around them, expanding outwards along the pathways. These myelin segments along the pathways sit alongside tiny gaps known as nodes of Ranvier, and it is these nodes that relay electrical impulses, node to node.[62]

As each node of Ranvier becomes squeezed more tightly by the adjoining myelin segments, it takes less time to trigger an impulse, and hence impulses are initiated more quickly. This means that pathways that are more myelinated transmit signals much more quickly than unmyelinated or less myelinated pathways.

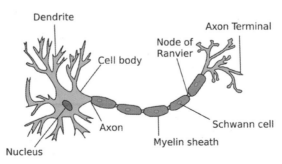

Myelin sheath

Learn, Repeat, Rest

Like any successful athlete will tell you, there is no substitute for putting in the hard work at the gym. Similarly, our brains – our myelin to be precise – need to exercise to get strong and fit.

Firstly, *learn*.

Microsoft CEO Satya Nadella once said: "Many who know me say I am also defined by my curiosity and thirst for learning. I buy more books than I can finish. I sign up for more online courses than I can complete.

"I fundamentally believe that if you are not learning new things, you stop doing great and useful things."

This fundamental thirst for learning and aspiration towards doing more is reflected in Nadella's leadership style and vision. In his first email to Microsoft employees as CEO, he said the reason he was in Microsoft was to change the world through technology that empowers people to do amazing things.

"I know it can sound hyperbolic – and yet it's true. We have done it, we're doing it today, and we are the team that will do it again.

"This is a software-powered world. It will better connect us to our friends and families and help us see, express, and share our world in ways never before possible. It will enable businesses to engage customers in more meaningful ways... we have unparalleled capability to make an impact."[63]

Without learning and being open to new perspectives and ways of doing things, working towards such a vision would be near impossible.

Next, *repeat*.

Observe a young child learning to button his shirt and you will see him struggle with it. Over and over again, he will try to fit the button through the buttonhole until it becomes easier, and finally, effortless. This is what happens with myelination. When we learn something new, neurons fire signals to take certain actions. The process is tedious and difficult, and likely uncomfortable, especially if it requires constant retrieval of information. It may even give you a headache!

Repetition activates neurons, which in turn grows myelin and helps hardwire the pattern of instructions.

A tip: Repeating a process at one sitting and in an unfocused manner can actually take you longer to learn and assimilate the process. Choose to break the process down into small chunks, and practice those chunks with more repetitions, instead of repeating the whole process fewer times. This allows for more repetition and more hardwiring of the process. Go slow, and build up speed so that your neurons get used to firing signals in the correct sequence, while making fewer mistakes (that could potentially wrongly hardwire!).[64]

Finally, *sleep*.

Related to frequent and short repetitions of our learning is how our brains assimilate new information. Breaks and sleep allow for the activation of neurons to be spaced out, rest, and in turn transmit signals more easily to different neurons.

Did you also know that most myelin production happens at night? Countless studies have pointed to the importance of sleep for brain health. The brain is powered by blood sugar and contains a limited amount of glycogen, which it uses up during waking hours. Sleep allows us to rebuild these glycogen stores, so our brains are recharged with energy at night. Without enough sleep, our brains run at low power and in the long term, cannot learn, remember and concentrate as well.

Your mom was right when she insisted you get your eight hours of sleep every night!

Stay Focused to Myelinate Well

There is a story about a client I once coached.

Jack (name changed) was a wealthy Asian businessman running his family conglomerate, and he approached me one day to ask if I could personally tutor him through his masters degree from a top business school in the United States. I moved temporarily to the United States, and worked with Jack as he learnt from the best and brightest in the country, including Nobel laureates in the faculty.

However, I noticed something about the way Jack was approaching his studies. Because of his business, his attention was constantly divided between his studies and work;

oftentimes with his attention very much more focused on his business through phone calls, instant messaging apps and email. It came as no surprise when the results were released and he barely passed the exams despite having access to the top brains and a personal tutor.

I realised that he had been myelinating the wrong neural pathways. Jack had been training his brain to constantly switch attention between studies and work, firing to different brain circuits simultaneously and haphazardly. It probably caused myelination in unintended brain circuits.

It was a clear indication that myelination is an extremely powerful tool that we have in our arsenal to build our brains for smartness.

So the next time you think about multi-tasking, and constantly at that, consider whether you are myelinating the intended neural pathways.

Beware of Synaptic Pruning

A note of caution though.

With all our attempts to build and strengthen our myelin, they will produce no fruit if we do not first protect the existing synapses that we have. After all, if there are no synapses, there is no need for myelination.

The world we live in has made it increasingly easy for a phenomenon called synaptic pruning to take place. To understand what this means, let us look at the analogy of a city with roads running through it, interwoven among buildings and other structures. If, after some time, the city council finds that a particular road is no longer in use because people find other roads more useful for their needs, they would eventually close up that road and use the space for other more pressing purposes. The city represents our brain, and the roads, our synapses. For every neural pathway that is no longer used, the efficient brain would naturally prune it away and use its precious resources in other ways that are deemed more valuable. The more synapses are pruned, the fewer connections there are in our brains, the less myelin is needed. This reduces the prowess of our brains.[65]

Add to the mix the digital traps of filter bubbles and echo chambers, plus digital addiction that we explored in Chapter 2, and we will find ourselves in a losing battle. Not only will synaptic pruning result in us forgetting things, these digital traps will slowly cultivate in us an inability to learn. We will gradually lose our learning capabilities, simply because our brains can no longer do it. The dangers of the digital world will be unleashed, a final blow to our brains.

Thus, we must protect our minds at all costs. How?

Ensure a variety of input to our brains. Experience new things often. One of the simplest solutions is to travel, in the wider sense of the word. It can be global travel, cross-country travel, or even travelling to different areas of your city and town that you do not normally visit. Eat at a different place from where you usually dine at. Walk different paths, visit different shops. Observe the differences in what you see, smell, hear. Allow your mind to wander. Take in what seems unusual, perhaps even uncomfortable. All this seemingly random and novel new input will find its way into the deepest recesses of your mind, ready as 'a-ha!' moments to be used in the most creative ways when needed in future.[66]

So, all hope is not lost. There is a way to become smarter and more innovative, to break out of the trap of System 1 and move into System 2.

CHAPTER FIVE

Escaping From System 1

Judo is one of the most-practised martial art forms in the world, with more than 50 million judokas around the world.

Founded in 1882 by Jigoro Kano, this martial art was introduced to the Olympic Games in 1964 and has since been recognised for its amazing throwing techniques, alongside grappling, arm locking and pinning moves. It is a combat sport that is swift, decisive and has spawned numerous other martial art forms, including the deadly Krav Maga, practised by the renowned Israeli defence forces.

But did you know that judo is also known as the 'gentle way' – *ju* from the ancient *jujutsu* to mean 'gentle', and *do* meaning 'path'?

When Kano first created this martial art form, he brought together a marriage of the principles of *ju yoku go o seisu* (softness controls hardness) and *seiryoku zen'yo* (maximum efficiency, minimum effort). The juxtaposition of gentle with dynamic means one is yielding and does not resist an opponent's moves, but it is in that seeming compliance that power is redirected back towards the opponent, resulting in his defeat.

When we think of strength, we associate it with attributes like hard, tough and non-compromising. Someone who is strong is powerful, rigid, an impenetrable fortress. To be strong is good, weak not so.

Yet contrary to common understanding, there lies great strength in the soft. As Bruce Lee famously said: "Be water, my friend." Water appears weak, but possesses within it much power; gentle forest streams can carve paths through rock over time.

Even in a traditionally hard and objective field like mathematics, it has been shown that taking the path of unknowns and uncertainty helps. In the 16th century, Italian mathematician Rafael Bombelli discovered that looking at numbers as abstractions, or imaginary numbers, rather than concrete entities, would allow some of the hardest problems on earth to be solved. Today, imaginary numbers are taught as elementary math, and are a fundamental cornerstone in many areas of mathematics and science.[67]

What has this got to do with escaping our System 1 way of thinking into the luminance of System 2? Surely we can easily will ourselves to slow down, think intentionally, engage our System 2?

I invite you to try a quick experiment in determining our ability to switch systems. Consider these brainteasers from a 2015 study in *The Journal of Problem Solving*:

1. A man is reading a book when the lights go off, but even though the room he is in is pitch-dark, the man goes on reading. How? (The book is not in an electronic format.)

2. A magician claimed to be able to throw a ping-pong ball so that it would go a short distance, come to a dead stop, and then reverse itself. He added that he would not accomplish this by bouncing the ball off any object, tying anything to it, or giving it spin. How could he perform this feat?

3. Two mothers and two daughters were fishing. They managed to catch one big fish, one small fish, and one fat fish. Since only three fish were caught, how is it possible that each woman caught her own fish?

4. Marsha and Marjorie were born on the same day of the same month of the same year to the same mother and the same father — yet they are not twins. How is that possible?[68]

(Answers: 1. The man did not require light to read because he was blind, and was reading the book in Braille. 2. The magician threw the ball upward into the air, not horizontally, so its motion was reversed by gravity and not by a collision with the ground, table or wall. 3. Only three fish were caught because there were only three women – a girl, her mother, and the mother's mother. 4. Marsha and Marjorie were not twins, but triplets.)

Did you manage to get any correct? On average, less than half of the subjects managed to solve the brainteasers. Their answers were based on preconceived frames of thought, and hence resulted in incorrect interpretations of the circumstances described by the riddles. It is natural that these preconceptions and interpretations come to mind automatically, with little or no conscious consideration. That is our System 1 at work. Even though we were aware of being tested, and had an inkling that we needed to engage our System 2, it was still difficult to break out of these norms of thought. Escaping from System 1 is hard work.

In order to escape this default prison of thought, we cannot depend on the usual forceful and hard methods. Not when we are dealing with subconscious cognitive processes. Often, the more difficult something is to do, the softer an approach is needed to find a way through it.

Nudging Towards Greatness

How we can free ourselves from this tricky trap starts from first understanding how human beings prefer to operate.

As human beings, we like our freedom. We like to be able to choose rather than be forced into something. Even modern parenting gurus expound on the benefits of letting one-year-old toddlers have choices over what to wear and play with.

We also like things that are cognitively easier to handle, a path of least resistance, so to speak. When things require way more brain power than what we think should be used, we often choose instead to skirt the situation or put it off, sometimes indefinitely.

Finally, we like things in manageable portions. It feels less threatening. That is all we can stomach before our brains go into overdrive and totally give up on the task at hand out of tiredness (or laziness!).

Against this backdrop of understanding the inherent irrationality of human beings, Richard Thaler, winner of the 2017 Nobel Prize, formed the nucleus of his work in behavioural sciences. Much like his compatriots, Daniel Kahneman and Amos Tversky, Thaler saw that something that may appear extremely logical and beneficial can be rejected by people for outwardly trivial reasons that are compounded by biases and emotions.

He developed what is now known as the Nudge Theory, made prominent by the book he co-wrote with Cass

Sunstein, *Nudge: Improving Decisions About Health, Wealth, and Happiness.* In it, the authors defined their concept thus:

"A nudge… is any aspect of the choice architecture that alters people's behavior in a predictable way without forbidding any options or significantly changing their economic incentives. To count as a mere nudge, the intervention must be easy and cheap to avoid. Nudges are not mandates."[69]

In simple terms, it refers to changing the context in ways that will encourage people to choose or behave in a certain way. Through what Thaler calls 'libertarian paternalism', choice architects can design environments that will allow people freedom to choose while also directing them towards what the better choices are. These ways are small, gentle little nudges that helps us move away from our autopilot System 1 method of decision-making and thinking towards the intended outcomes.

Nudges may appear small, but they are ripples that bring about big waves of change. From individuals to businesses, there is a global recognition of the power in something that appears so small and simple.

Even governments have gotten in on the act. In 2010, the Behavioural Insights Team – also referred to as the Nudge Unit – was set up by then-Prime Minister of U.K. David Cameron, with a simple aim of helping citizens make better decisions for themselves. The Nudge Unit was put to work influencing the macro policy-making in order that the micro decision-making on the ground would be steered in the intended directions. It was a case of thinking small in order to achieve big goals – and it certainly did, seeing success played out repeatedly across the country. Over in the U.S., President Barack Obama also employed the Nudge Theory in domestic policy-making; today almost all state leaders apply various forms of nudging to execute their countless policies.

If external nudging can work on a national level, can we also use it internally on an individual level to move out of our immediate responses arising from System 1?

We certainly can, and the rules are simple. There are essentially three of them to bear in mind when we decide what and how we can nudge ourselves away from the default way of thinking, into intentionality.

Rule #1: Choose the Highest Curiosity

When you want to solve a problem with an unknown solution that requires System 2, try to start from the most novel part of the problem.

Can you remember the last time you were curious about something and felt a sense of coming alive? Perhaps you found yourself noticing the sights, sounds and sensations around you, being attentive and mindful to the here and now. This natural curiosity is what will help send a signal to your System 2 to activate and purposefully interact with the task at hand. At the same time, a self-directed and self-led curiosity means you will want to stay engaged (remember the importance of freedom of choice!) with your System 2 for a longer time. It will become easier over time for System 2 to come to the fore.

Rule #2: Focus on a Single Goal

We have possibly all done this at some point in our lives. New Year's Day rolls around and we decide that it is the best time for a new beginning. We list the things we want improved, and resolve that this year will finally be the year we get fit, eat healthy, spend more time with the family, do good. By March, that list will likely have been stuffed into a corner of our mental drawer, because well, life got in the way. Too many other things to do. And the cycle repeats the following year.

The issue is this: When we try to work on several ambitious goals at the same time, our efforts are undermined. The cognitive effort required to achieve one goal will wear away at the effort needed to achieve the others. In other words, for most of us, the problem is not a lack of goals; it is having too many of them.[70]

Similarly, when we have a difficult or novel problem we need to solve, focus instead on a specific part of it. This approach of the simplification principle from nudging will help us coax our lazy System 2 to work longer at solving the problem.

Rule #3: Chunk Your Goal

Britain's national cycling team were a dismal lot for much of their history, until 2003, when Dave Brailsford was hired as the team's performance director. He looked for opportunities to make changes in every aspect of cycling – from bike seat design and racing suit fabric to managing the way they washed their hands, and even the pillows and mattresses they slept on. From mediocrity, they vaulted to 16 gold medals in the 2008 and 2021 Olympics and won seven Tour de France races over eight years.

The massive leap in improvement arose not from big changes, but from what Brailsford called the aggregation of marginal gains. Said Brailsford: "The whole principle came from the idea that if you broke down everything you could think of that goes into riding a bike, and then improve it by 1 percent, you will get a significant increase when you put them all together."[71]

Brailsford's marginal gains theory of 1% improvement is at its core the concept of chunking. Breaking things down into bite-sized pieces is easier to do – we feel less intimidated and more confident whenever we succeed at completing one part. Success then builds on success.

To chunk a problem, work out what the different components, stages or components are to the task at hand. Chunking can also come in the form of breaking down the overall objective into parts. Both ways allow us to apply the ease and convenience principles from nudging.

Keep in mind these three rules to nudge us out of System 1, because they will guide us on our path towards System 2. Over the rest of this chapter and the next two chapters, I will show you four steps to take on that path: cultivating a Quiet Brain, making Mistakes, performing Next Practice Rituals (R), and undergoing Next Practice Training (T).

The first, and really the basis of the way out, is the Quiet Brain. Creating a literal and figurative space for quietness – through activating our Default Mode Network for the work of creativity and imagination, and finding our flow – will naturally allow our Quiet Brain to take centrestage.

Let us take a closer look.

Mindful Flow of the Default Mode Network

It was under the soothing shade of apple trees that Isaac Newton, in a state of contemplation and with a cup of tea in hand, discovered the law of gravity. There, amidst the sweet scent of ripe apples, Newton observed the falling of yet another apple and wondered why.[72]

That almost dreamy wandering of his mind led to one of history's most well-known laws of physics.

It may seem surprising that such brilliance came forth in a seemingly serendipitous way.

The common sentiment is that in order to solve a problem or tricky situation, we need to think long and hard about it. Spend hours researching, digging for solutions, slogging away and stopping only when we have found the answers we need. Companies and government offices are full of expressions such as doing overtime, pulling all-nighters, and similar concepts that glorify the hardworking employee who works long hours. But the truth is, ideas are often born out of the playground of idle minds, not busy ones.

Neuroscience can give us an insight into this anomaly.

Our brains contain many different networks, one of which is known as the Default Mode Network (DMN). As nondescript as the name sounds, it is actually where much of our creativity arises.

Recall the last time you had a sudden insight into something that you had been mulling over for a while. Chances are that insight came about when you were doing something really mundane – perhaps driving, showering, or even while taking a walk. New ideas bubbled up with an exuberance. Things started to make sense when previously they may not have.

That is exactly how our DMN works, in a counter-intuitive manner. The beauty of the DMN is that it is active when we are not consciously thinking about something, when we are daydreaming or just allowing our minds to wander and spontaneously think. It is triggered by default when our brains are not paying attention to anything in particular.[73] When that happens, our DMN is abuzz with activity. It turns away from the many stimuli we have been focusing on, and looks inward to start joining the dots. It may go back to the past and recall memories, into the future to paint a vision, consider what someone meant when they said something, scan the surrounding environment to notice quirky details – basically things that we randomly do when we are just thinking without any explicit goal of thinking in mind.[74]

 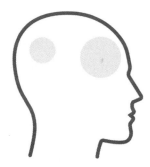

During leisure tasks (watching TV, socialising, travelling, reading, etc.), the **DMN is activated**

During cognitive tasks (retrieving information, performing calculations, etc.), the **DMN de-activates**

Because the DMN resides within sub-regions of the brain called the association cortices, it is there that the DMN starts to access the database of memories and feelings stored in the rest of our brains, and make associations we would normally not make. These associations help us create meaning; they are the source of our inventiveness.[75]

Creativity does not come from nowhere, ideas do not come out of nothing. We often mistake ideas as random serendipitous happenings that are plucked out of thin air, or we think that the domain of new ideas belongs to people blessed with a creative streak. Not at all. Creativity is often a result of our brain's wonderful ability to make associations, combine concepts and fill in the gaps.

In a nutshell, creativity is the brainchild of our imagination.

Imagination Moulds Reality

What we can imagine can become reality.

For something to be created, someone first had to imagine it. According to historian Yuval Noah Harari in his book *Sapiens: A Brief History of Humankind*, the key difference between humans and animals is our ability to imagine things, and then collectively treat them as a part of reality.

Something exists because someone imagined it, and as a community, society, human race, we chose to believe it and create systems that aligned with those beliefs. The understanding that the world is round, not flat. Aeroplanes that transport us around the world. Computer systems and cloud computing. All of these, and more, would never have become reality if they did not first start from the imagination.

It stands, therefore, that our imagination moulds reality. Anything that we can imagine, and can convince others to support, can become a concrete reality. The imagination is a tool far more powerful than any other skill of our brain. It is where knowledge and creativity are harvested, from which civilisations are made.[76]

Finding Flow

According to Dr Nancy Andreasen, who has spent decades in the field of neuropsychiatry, "When your mind is at rest, what it is really doing is bouncing thoughts back and forth. Your association cortices are always running in the background, but when you are not focused on some task – for example, when you are doing something mindless, like washing your hair – that's when your mind is most free to roam. That's why that is when you most actively create new ideas."[77]

And this is where we want to be if we hope to tap into the immense power of our brains – the imaginative power that births ideas, drives innovation and brings us into that sweet spot called *flow*.

Inventors experience it. So do jazz musicians, writers, and anyone else who knows what it is like to enter an almost hypnotic state where time seems suspended. It is a perfect alignment of senses, thought and action. There is almost no more stress about what and how to do something; it happens effortlessly with no tension of a beginning or end. Being in a state of flow is being in a space where everything feels free, complete and seamless.

This is flow. In this place, imagination bursts forth. It is an optimal state of consciousness.

By scientific definition, it is the state of transient hypofrontality that enables the temporary suppression of the analytical and meta-conscious capacities of the explicit system.[78]

So we know that flow happens when we are in our DMN. And in order to activate our DMN, the key lies in resting our busy mind.

The resting brain is not inactive. In fact, the resting brain is infinitely more active than we realise.

Taming the Noisy Monkeys

What better way than to tap into the centuries-old practice of meditation?

Hesychasm, or Greek for 'stillness, rest and silence', is a mystical tradition originating from the Christian desert fathers that was central to their practice of interior silence and prayer. Zen, coming from a Sanskrit root meaning 'thought' or 'absorption', holds meditation at the heart of its practice, with an emphasis on insight and self-control. In both these meditation practices, and those of other religions and practices, the focus has always been on quieting the thoughts in one's mind.

As New Age as it may sound, meditation has been scientifically proven to enhance imagination, opening doors where once there seemed to be impenetrable walls.[19]

In a study done on the effects of meditation on the brain's connectivity among and within networks, the findings demonstrated that when meditation is practised regularly, it increased the brain's ability to switch between the DMN and the ability to focus, and maintaining attention once in the attentive state. Moreover, structural magnetic resonance imaging (sMRI) found that meditation increased the density of myelin and reduced perceived stress. The study found that participants of a 2-month meditation training showed increased brain connectivity even when they were not in a meditative state.[80]

Simply put, meditation has the power to increase our brain's ability to move from the movement and attentiveness of our busy minds to the stillness of our DMN in System 2.

I always suggest to my students and clients to do a walking meditation when they get stuck trying to solve a wicked problem. The practice is simple yet immensely powerful. Walk slowly on a short path back and forth, deliberately, so that you are never out of breath, and on a path familiar enough where you will not be distracted. Walking meditation removes the burden of thinking and distraction, and allows you to be aware of the present moment. There is no hurry, nothing to get to. In doing so, your DMN will awaken and work its magic.[81]

Idle Minds Come Alive

The author of *Elastic: Flexible Thinking in a Time of Change*, Leonard Mlodinow, describes a series of conversations he had with Professor Stephen Hawking where he had to wait for a long time in between each question for Hawking to type out his answers. While he was initially bored, he soon realised that "the stretching of those seconds to minutes had a hugely beneficial effect. It allowed me to consider his remarks more profoundly, and it enabled my own ideas, and my reaction to his, to percolate as they never can in ordinary conversations. As a result, the slowed pace endowed my exchanges with a depth of thought not possible in the rush of normal communication."[82]

The power of relaxed idleness was also uncovered by Paul Lewicki of the University of Tulsa and his team. A group of subjects was asked to pay attention to a computer screen that was divided into four quadrants. An X would appear in one of the quadrants, and the participants' task was to press a button predicting which quadrant the X would next appear in. Unbeknownst to the subjects, that order was dictated by a complicated set of rules. Despite the complexity, subjects were soon able to guess where the X would appear after the initial mistakes.[83]

The conclusion was that the participants, who Lewicki always ensured were in a relaxed state of mind, aced the experiment because their minds were able to access their DMN and learn the complicated pattern without even realising it. They became faster over time at pressing the correct button, and when the rules suddenly changed, their performance dropped badly, before picking up again.

Having a relaxed mind is what gave the subjects a superior ability to 'predict' what the next move would be, filling in the gaps where the analytical mind would not have been able to, because the rules were way too complicated.

The importance of relaxation is one that cannot be emphasised enough.

Yet relaxation to the point of being aimless may not sit well with a culture that values productivity and efficiency. We consider ourselves productive and useful when we are constantly rushing and virtually active 24/7. Being busy is a badge of honour we flaunt with pride as a symbol of being productive. Being relaxed belongs to unproductive people.

Says Mlodinow in his book: "The result of our addiction to constant activity is a dearth of idle time, and hence, a dearth of time in which the brain is in its DMN… [Downtime] allows our integrative thinking processes to reconcile diverse ideas without censorship from the executive brain.

"[Internal conversations of the brain] allow us to connect divergent information to form new associations, and to step back from our issues and problems to change the way we frame them, or to generate new ideas… to search for creative, unexpected solutions to tough problems.

"The associative processes of DMN do not thrive when the conscious mind is in a focused state. A relaxed mind explores novel ideas; an occupied mind searches for the most familiar ideas, which are usually the least interesting. Unfortunately, as our DMN are sidelined more and more, we have less unfocused time for our extended internal dialogue to proceed. As a result, we have diminished opportunity to string together those random associations that lead to new ideas and realizations."[84]

The next time you find yourself in a sticky situation where you cannot find a solution, find a way to relax. Get some fresh air, stare out of the window, do some stretching, listen to music, talk to a pal, do aromatherapy. Even a pause to have a drink of water can help provide micro relaxation amidst the barrage of thoughts and ruminations. (Or take a look at the list of 81 Relaxation Techniques in the Appendix.)

My specific advice to students and clients searching for their Next Practice is this: If all your conscious effort to prepare, study, and find solutions ends up in you hitting the wall numerous times, it is a sign that your unconscious mind, or DMN, is finally ready to take over. Just stop. Walk away, speak to your DMN to come to the fore, and then find a way to completely relax.

Working until we feel the struggle is the key to activating our DMN. Struggle is essential. Getting stuck is needed. Until that point, the DMN will remain dormant, no matter how relaxed we are.[85]

Learning and repetition entail struggles. Stay with it. Persevere and push through, and you will find yourself struggling less, simply because the neural pathways have formed and are being strengthened.

This is why procrastination has its merits (vindication for the procrastinators of the world!). Data shows a positive correlation between procrastination and imagination, because putting off attempts to solve problems and make decisions gives our DMN time to percolate the possible solutions on the backburner.

Leonardo da Vinci was known to take frequent hiatuses while working on *The Last Supper*. As the art historian Giorgio Vasari recorded, "The prior of the church entreated Leonardo with tiresome persistence to complete the work, since it seemed strange to him to see how Leonardo sometimes passed half a day at a time lost in thought, and he would have preferred Leonardo, just like the labourers hoeing in the garden, never to have laid down his brush."

Leonardo's reply? That the greatest geniuses sometimes accomplish more when they work less – i.e., relax.[86]

And we know how that masterpiece turned out.

CHAPTER SIX

Mistakes: The Master Key to Innovation

Failure is good.

We should seek it, encourage it, reward it. Give employees bonuses for making mistakes, students a pat on the back for messing up yet another assignment, government officials credit for flaws in their policies.

Now pause.

How did that make you feel?

If you are like the average person on the street, reading it would possibly have made you somewhat uncomfortable. Although it has become trendy in recent years to embrace making mistakes as part of learning, the reality is that systemically, society does not support failure. There exists an almost leprous stigma to it. It is still the students who ace their exams who get places in the top schools. Tools that weed out probabilities of error, such as Six Sigma, are still widely used by businesses seeking to be best in their field. And woe be on the engineer or doctor or lawyer who makes a mistake in their area of work, because the stakes can be high. So everyone buckles down to minimise mistakes and risks in order to avoid failure. No one wants to be remembered as The One Who Made a Mistake.

In the end, to say we welcome failure is mere lip service. After all, it makes no sense. Wouldn't encouraging failure result in mediocrity, slipshod efforts and an overall lowering of standards, standards that human beings have for generations fought hard to raise for the betterment of their lives and society in general?

Failure has been a bad word for a long time, seen as an antithesis of success.

It has been unfairly shamed – and the time has come for it to take its rightful place in the spotlight.

Why is Being Wrong, Wrong?

Every time I meet a class for the first time, I open it with just one rule.

Make mistakes.

My students, who often wait with trepidation to hear it (no one likes rules!), are relieved and laugh nervously, but are puzzled. None of their other professors set such rules. But for me, I value a student who is able to make mistakes and learn from them, because the work that comes after is often of higher quality, well thought-through, and more robust.

In the book *Art and Fear* by David Bayles and Ted Orland, there is a story about a ceramics teacher who told his class that half the class would be graded on the quantity of pots they produced – an A for 50 lb of pots, a B for 40 lb of pots, and so on. The other half would be graded on the quality of the pot they produced; yes, they only needed to bring along their best, perfectly crafted piece.

The result? The most creative, well-made, high-quality pieces all came from the group graded for quantity. Bayles and Orland commented: "It seems that while the 'quantity' group was busily churning out piles of work – and learning from their mistakes – the 'quality' group had sat theorising about perfection, and in the end had little more to show for their efforts than grandiose theories and a pile of dead clay."[87]

The process of innovation, and Next Practice, is truly an exercise in letting go of control to get things right. There will be many moments when the path ahead is unclear, attempts to find the way out unsuccessful. Two-time Nobel laureate Linus Pauling rightly puts it: "The way to get good ideas is to get lots of ideas and throw out the bad ones."[88]

It is hard to let go; failure is an incredibly uncomfortable experience. Yet, mistakes liberate us from the hold that our System 1 has on us. They reduce our overconfidence in the limits of our knowledge, allowing us to move past our illusion of understanding to see what we still do not know.

Furthermore, because of our tendency to seek evidence to confirm what we know, we often do not see the alternatives, and remain blinded because of the illusion of validity. Mistakes are necessary.[89]

Surely it must come as a relief to know that mistakes can — and in fact, should — be welcomed. Being allowed to make mistakes provides a sense of liberation from fear, and opens the way to being uninhibited. A great place to observe this is to look to young children, and how they are fearless in their thoughts and questions. Their elasticity of thought very often leads to an uncensored flood of ideas, and from those come marvellously creative possibilities. If only we give ourselves permission to make mistakes as well!

Consider this: At the back of our minds, we fear mistakes because of the potential consequences. What if my recommendation turns out wrong and causes my company to lose millions of dollars? What if taking a year off to travel the world causes my career to derail and I lose the love of my life waiting at home for me? What if my decision to study theatre results in me being a penniless and struggling artist? There are so many what-ifs, and even more consequences of catastrophic proportions that we imagine in our minds. They stop us in our tracks, and we tiptoe ever so carefully and hesitantly to avoid getting hurt.

However, we forget that in reality, we make mistakes that span an entire spectrum, from inconsequential ones to those that carry with them serious consequences. And the dramatic irreversible ones we imagine? They are a minority, possibly making up just 5% of all mistakes. The remaining 95% are mistakes that help us get better and grow stronger. Yet the 5% is what we keep our eyes on, and it holds us hostage, preventing an unbridled spirit of trying, for fear of failing and falling hard.

We need that 95% if we are to discover the Next Practice. We need to switch our focus.

Fail, to Succeed

Chess is a game of strategy familiar to many. In order to win, players must capture their opponent's king in a sometimes complex combination of moves, moves that chess grandmasters are adept at manoeuvring to retain their titles. Not even the smartest of computers had been able to defeat these experts.

Until Deep Blue proved the world wrong.

In 1997, the IBM supercomputer challenged then World Chess Champion, Russian master Garry Kasparov, and beat him in a historic match. It was able to evaluate a staggering 200 million board positions per second, get the right answer and predict Kasparov's moves, eventually taking him down. It shook the belief that human beings were inherently more intelligent than computers.

Even with that stunning victory, there was one other game that continued to elude the best computer programs in claiming final victory over human beings. The game of Go.

The rules of Go are simpler than those of chess, yet its complexity is much, much greater. No computer could cover all the possible moves that would result in a winning formula – it was a number beyond imagination and completely unrealistic to exhaustively evaluate.[90]

The only way for a computer to win was to make educated guesses based on probability, which brought with it the chance of making the wrong guesses. Which is exactly what computer software AlphaGo did in 2015, beating Fan Hui, winner of the European Go Championships.

Instead of Deep Blue's method of exhausting all options to find the right strategy, AlphaGo – built not on sophisticated machinery but on off-the-shelf processors – went with picking the most promising series of moves and evaluating them to the end, in what is known as a pseudo-random sampling, or Monte Carlo tree search. It then went on to play millions of games against itself, learning and then improving upon its performance in the process.

Put simply, AlphaGo had to guess and make mistakes, and do that over and over again, in order to succeed. It did not aim to conquer vast amounts of information and climb mountains of unknowns; rather, it came out on top just by picking the easier options with a lower burden of knowledge, and exploring them. Much like Next Practice, AlphaGo stood apart from the rest because it used mistakes as a springboard to reach far greater heights in innovating.

Mistakes are the hidden gems that many do not see.

Science teacher Anne Smith teaches at Carmel Catholic High School in Mundelein, Illinois. Her 9th grade physics students are in the midst of learning about electric circuits.

Armed with an array of tools such as paper clips, batteries, tape and a lightbulb, they are let loose to try and figure out how to work an electric circuit. "Have a go," Smith says to them. "See if you can make the bulb light up."

A strong believer in the power of trial and error, she shares: "When students are allowed to work through difficult material, they gain confidence. They learn that making mistakes is part of the scientific process."

It is a refreshing viewpoint also held by neurobiologist Stuart Firestein, author of *Ignorance: How It Drives Science*. To him, failure is one of the two most important ingredients in science, together with doubt.

"When an experiment fails or doesn't work out the way you expected, it tells you there is something you didn't know," he says. It suggests a need to reverse the steps taken to examine what went wrong, and why.

In the deep search into what he calls "the portal of the unknown", the most valuable questions come, the kind that catalyse new ideas and types of experiments. And when a scientist discovers a new or better question, "that is where the action is. [Failure] propels science forward."[91]

In fact, a study in 2011 by psychologist Jason Moser and a team of researchers at Michigan State University showed that brain activity rose when a mistake was made.

Twenty-five participants, wired up to machines while completing a test with 480 questions, provided data that those who gave more thought to questions they got wrong did better on the test subsequently.

"When a participant experienced conflict between a correct response and an error, the brain was challenged," Moser reported. "Trying to make sense of this new knowledge was a time of effort and need for change... [The conclusion is that] by thinking about what we got wrong, we learn how to get it right."

"Two-thirds of Nobel laureates have announced their winning discovery was the result of a failed experiment," says Firestein. Penicillin, X-rays and insulin came about because of failed experiments. Thomas Edison failed more than 9,000 times experimenting on alkaline batteries (yes, he didn't only invent the light bulb!) before he patented the winning one.

Legendary basketball star Michael Jordan, too, credits failure for his success. "I've missed more than 9,000 shots in my career. I've lost almost 300 games. Twenty-six times, I've been trusted to take the game-winning shot – and missed. I've failed over and over and over again in my life. And that is why I succeed."[92]

As Albert Einstein wisely said: "The only sure way to avoid making mistakes is to have no new ideas."

For if we never make mistakes, never have a chance to learn from them and grow in wisdom, we will never reach our dreams. It is a constant spiral trajectory that our lives must take to get better. Like *yin* and *yang* that are natural bedfellows, so too do failure and success possess a synergistic complementarity where neither can exist without the other.

I often tell my students never to fall in love with their ideas, unless they wish to get hurt. As most of us have experienced, there is an irrationality that comes with falling in love. That same irrationality can also affect us when it comes to the intellectual realm. Love is blind, and it is a very dangerous variable to have if we apply it to our ideas as well, as demonstrated by Kahneman's illusions of validity and understanding. The instant we get too attached to our ideas, we fail to stay objective, and will be unable to detach our ideas and opinions from ourselves, our identity. Our ideas and opinions arc us. This will become a problem when it prevents us from changing our minds, creating new ideas, as the world changes and knowledge evolves alongside.[93]

It is only when we can see our ideas and opinions as separate from our identity, that we feel the freedom to create, fail and create, over and over again, until we succeed.

The Next Practice requires us to take this same attitude and willingness to try unceasingly, until we find what we are looking for. It is clear that failure is a must, not an option.

Without failure, success cannot happen.

Wild Guessing Our Way

By definition, Next Practice is the creation of an avant-garde course of action taken to solve critical challenges and problems faced by an organisation. Being avant-garde is a key characteristic of the search for the Next Practice; it favours new and experimental ideas and methods.

Experimentation today is no longer the domain of scientists in laboratories. Businesses across various industries are cottoning on to the benefits of experimentation. Experiments spur innovation because of their built-in allowance for correction of faulty intuition, inaccurate assumptions or overconfidence.[94]

One way experimentation can be done is through the use of approximation.

Approximation is a way of guesstimating an answer – a kind of 'back of napkin' method of roughly coming up with a solution. It often turns out wrong, but this way of calibrating is incredibly useful in leading us close enough to the right answer.

Imagine being in a rather large room that is pitch-dark. There is only one exit and the task is to find it, fast. What can be done? Some may systematically feel their way along the walls, inch by inch, which will probably take a very long time. Others may instead just try their luck and walk, adjusting their route based on what they already know about space, dimensions, echolocation, etc., and perhaps achieve their goal quicker. The latter is what we call approximation.

Google is a company with innovation in its DNA. Innovation is crucial for its survival, and it needs employees who are similarly built for innovation. So for over a decade, Google has been known to conduct job interviews that ask questions which have no answers, in order to suss out the thought process behind a candidate's answer. Questions like "How much does the Empire State Building weigh?" and "How many basketballs will fit into a city bus?" have been tackled in a variety of ways, but one of the most useful methods successful candidates use is figuring out an approximate answer. Because there are no right answers, these candidates exercise their ingenuity to find connections between chunks of information they already know, experiment with possible hypotheses, combinations and links, before they solve the problem. This way of critical thinking brings together a reasoned and organised mind with imagination and a sense of exploration in thought and ideas – all a fertile bed for the Next Practice to emerge.

If we think about some of the greatest inventors, they were not the first ones to invent the thing that propelled them to fame. The Wright brothers did not invent the first aircraft that could take flight, but they worked on numerous prototypes before they created the aeroplane that we all recognise today. Neither was Thomas Edison the first inventor of the light bulb. What he did do with his team was to experiment on hundreds of materials before finding the winner. Put simply, Edison succeeded because he was willing to keep experimenting. So were the other great inventors. They just kept trying.

Many great thinkers frequently turn to another classic form of approximation, known as thought experiments. When we encounter seemingly impossible situations, using thought experiments gives us a chance to rhetorically consider implications and outcomes, without needing to set up an actual experiment.

If we want to seek the Next Practice, approximating in this manner opens our minds to speculation, logical thinking and paradigm changes all at once. We are no longer held back by norms of what can or cannot be done, and see instead possibilities.

Can we move out of our comfort zones to face situations we do not have answers to? Are we able to sit with unknowns and acknowledge that we do not know everything?[95]

Ambling Into New Ideas

It is a familiar magic that happens for some of us when we visit a finely curated bookshop.

Like the children who entered the wardrobe and chanced upon the magical land of Narnia, we go in wanting to find a particular book – and wander into a whole new world. We end up meandering through the aisles, our attention caught by a titillating word, an insightful title, a bold cover that demands to be looked at. Time is suspended as our consciousness explodes with bursts of delight at new discoveries.

I thoroughly enjoy dropping in at Books Kinokuniya, a book store in metropolitan Singapore. It is a book store like no other. Away from the bustle of the nation's main shopping belt, the shop is an Aladdin's cave of the best and latest titles from all around the world. Yet its uniqueness lies not in its comprehensive list of books, but in its flair for sifting through different book genres and bringing together titles that work well together, with great intentionality.

As a reader browsing the myriad of tiles, I am able to pause, reflect, and get pulled further and deeper in as countless ideas in the books spark off connections and links I would not have made, had I simply gone into Kinokuniya, taken the book I was there for, and left.

The chance to randomly browse, and find new connections, is a privilege that is slowly getting scarce in the technologically driven ecosystem we exist in currently. Instead of walking down aisles and perusing tables of books, we home in on Google search with laser focus, get the information we need, and click away. There is no more accidental chancing upon an unrelated idea, article or book. We lose the happy cognitive buzz we get from the spatial and tactile stimulation of holding a physical book, flipping through its pages; this is something that digital-format words – even in electronic books – can never give us.

We must bear in mind that the process of finding the Next Practice is much like the joy of leisurely browsing in a book store or library. Giving our minds the space to roam free allows it access to a kaleidoscope of opportunities for ideation to take place – the perfect breeding ground for Next Practice.

Failure Brings Opportunity

Our ability to make mistakes is our pièce-de-résistance in the ongoing discourse to prove that human beings will never be taken over by computers.

Computers can do amazing things. Artificial intelligence comes close in being able to predict our actions and decisions, even drive our cars and beat the top grandmasters in chess. But what they lack is the ability to know when they do not know something, and they continue behaving as they have been programmed to do. That false sense of confidence can eventually cause their downfall.[96]

When we are able to guess, to know that we are uncertain and question ourselves – that is when we are able to pursue greater learning, and explore more options and variables. Clearing those unknowns will come about when we are willing to accept potentially making mistakes, for it is when we can face up to the need for failure that we will taste the true success of the Next Practice.

Mistakes as Way to Next Practice

My clients and students often ask me: "Andreas, if we are ready to make mistakes, what kind of mistakes should we be making in order to create the Next Practice?"

They are right to ask that question. We cannot assume we have carte blanche in making just about any mistake, with no boundaries to help direct us. If I am a doctor, and I fail to properly administer an injection, that surely is not an acceptable mistake. We must be able to at least perform the most basic tasks required in our roles.

There are two types of mistakes that you should make to discover your Next Practice.

1. Picking the right trends

If we think about it, ideas are aplenty. The issue is finding the right idea to capitalise on. As counter-intuitive as it seems, the quickest way is to scan the landscape for as many ideas as possible that have a low burden of knowledge. These are the ones that present themselves to us as weak trends, unlikely ideas, trends that are on the periphery of what is the latest and most exciting – those ironically are probably already the ones with a higher burden of knowledge.

Consider Rohit Bhargava, a 'Trend Curator' at Microsoft. He scours countless sources of content – conferences, magazines, periodicals, online articles, conversations, interviews with associates, etc. – and using what he calls his "haystack method", Bhargava sorts and sifts and shifts the material he and his team have found, ripping out physical pieces of paper with words and ideas that stand out and using Post-Its to note ideas. Gradually, connections are made, combinations arise, synchronicities emerge and trends appear.

"You've got to look somewhere other than where everyone else is looking," said Bhargava. Be fickle and keep moving through different types of ideas. Be observant and curious. "Curation is the ultimate method for transforming noise into meaning."[97]

2. Applying an innovation framework

There are many innovation frameworks available, such as Design Thinking and First Principles, to name a few. Use one of these to guide you to make educated mistakes. The innovation frameworks have been designed to ensure that even when we make mistakes, they are valuable ones.

**"To make mistakes is human,
but to profit by them is divine."**
Elbert Hubbard

In a 2018 paper published by Daniella Kupor of Boston University and Kristin Laurin of the University of British Columbia, they wrote: "People often fear that others will judge them as less likely to successfully achieve a goal if they make a mistake during their pursuit of it." However, their research shows that making a mistake (and fixing it) can actually play to one's benefit.[98] Humans can profit more from mistakes when we are able to correct them, rather than simply avoiding them.

We need to genuinely, and urgently, reframe our perspective on failure. Mistakes are requisites for shattering our illusion of validity and understanding. Our confirmation biases and overconfidence can stand in the way as well. We must embrace a culture of mistake-making in creating our Next Practice.

CHAPTER SEVEN

Next Practice Ritual (R) and Training (T)

If you have ever watched the New Zealand All Blacks perform the iconic haka, chances are you would have been left with a sense of awe – and feeling rather intimidated.

It is a stunning show of power and strength. The Kiwi rugby players face their opponents, and aggressively stamp their feet, beat their chests, grunt and shout, and use the most fearsome facial expressions they can muster. It is almost a challenge to their opponents to stay away or face defeat, much like what was originally intended as a war dance performed when the indigenous Maori tribes met in battle.

Since 1888, the All Blacks have performed this as part of their identity. It is a dance meant to scare the enemy away and rouse their own morale to win the match. For the All

Blacks, this is a pre-game ritual they never do without at every match.

A ritual is a mysterious thing. On the surface, it looks like a habit or a routine. You do an action repeatedly, sometimes with little effort put to thinking about it, and it may even seem like it is done on auto-pilot.

But unlike a habit or a routine, a ritual is something we do that has a meaning or symbolism attached to it. The meaning may not be a logical consequence or result of that action, but we believe it anyway. When we engage in a ritual, we become aware of something special happening, and have a heightened experience of it, turning us away from thoughtlessness towards a larger meaning.[99]

Athletes are famous for going through rituals pre-, during and post-game, such as basketball legend Michael Jordan's powder clap and tennis superstar Rafael Nadal's fixation with his water bottles and how they are aligned. Students are known to put their faith in favourite stationery when taking exams, and it is not uncommon to hear of people insisting on using particular clothing when they attend important meetings.

These actions and items do not make sense; they may even seem like superstitions. Yet, the meaning behind them trains our thoughts and subsequent actions, much like how the All Blacks do the haka as a ritual to prepare themselves (and their opponents) for the battle that will take place soon after.

Similarly, rituals form an important part in preparing us to enter the art and science of Next Practice. When done repeatedly, rituals create a psychological space that prepares our minds to start viewing things from a new perspective, to have an openness to possibilities that were previously blind spots.

We should engage in these rituals in order to activate our System 2 and find our Next Practice. These are age-old rituals that already exist, yet have never been lucidly explained by successful practitioners. It is precisely why innovation has been such an elusive practice – until now.

Holding Space in Quiet

"Solitude is a crucial and under-rated ingredient for creativity," says Susan Cain, author of the bestseller, *Quiet: The Power of Introverts in a World That Can't Stop Talking*. "From Darwin to Picasso to Dr Seuss, our greatest thinkers have often worked in solitude."

Solitude has a remarkable power to unlock our creativity and imagination. Free from the typical demands of people and the nitty-gritty of our daily routine, our minds get a chance to slow down and move at its own pace, unhindered. Solitude helps our brains relax, block out the external world and retreat inwards where the inner critic is silenced and our subconscious becomes free to incubate ideas. It essentially activates our DMN.

Being alone also frees us from something known as the

Spotlight Effect. When we are among other people, we tend to be conscious of them and their views of us. It is only when we are totally alone that we let our guard down, since there is no more need to keep up with what we think people think and expect of us.[100] Even though human beings are generally social creatures at heart and fear loneliness, solitude is not the same thing. Solitude is liberating. It is mental, psychological space so we can be at ease.

To understand the immense power of solitude, we need only turn to Microsoft's founder Bill Gates and his ritual of taking what he calls a Think Week. Once or twice a year, he arrives by helicopter or seaplane at a cabin in a cedar forest in the Pacific Northwest. Neither family nor friends are allowed to visit; he is completely alone save for someone who drops off two meals every day.

Armed with a steady supply of Diet Coke and Diet Orange Crush, Gates will spend the week poring through papers written by Microsoft employees pitching potential investments and new innovations. He puts in as many as 18 hours a day, reading as many papers as possible, making copious notes on them and doing his own deep thinking and ideation.[101]

Said Melinda Gates of her ex-husband: "Bill can deal with a lot of complexity. He likes complexity and he thrives on complexity. So when Bill stills and quiets himself, all these incredibly complex thoughts that he's had... he can pull ideas together that other people can't see, he thinks his best.

Why did he go on Think Week? He stills himself, and he has time to distill and slow down, and then write, and lead in the way he wanted to lead."

Think Week is a ritual that Gates has kept to since the 1980s, and it has spawned some of the biggest and best ideas, such as his 1995 paper, 'The Internet Tidal Wave', which led to Microsoft developing Internet Explorer, which demolished Netscape.[102] By the end of a Think Week, Gates would have blazed through as many as 100 to 120 papers, written a Think Week summary for executives, and sent emails to hundreds of staff for them to follow up on. Investments, changes in directions on plans, new company acquisitions kick into action globally, all of these a result of Think Week.[103]

The physical isolation that Gates puts himself in affords him a disconnection from family, co-workers, technology and everything else that places demands on him, in order to gain the mental space to create and focus.

Said Alex Soojung-Kim Pang, author of *Rest: Why You Get More Done When You Work Less*: "When we are mind-wandering, when our minds don't have any particular thing they have to focus on, our brains are pretty darn active. When you do things like go for a long walk, your subconscious mind keeps working on problems. The experience of having the mind slightly relaxed allows it to explore different combinations of ideas, to test out different solutions. And then once it has arrived at one that looks promising, that is what pops into your head as an Aha! moment."[104]

Like Gates, we too must make time for our own Solitude Moments as a ritual for discovering our Next Practice. Think of it as an extended shower. We often have flashes of brilliant ideas and insights while showering, yet these fade almost as soon as we step out. With longer Solitude Moments that we create for ourselves, it gives us an extended period of time for ideas to percolate, connections to be made among seemingly random things.

Find a third space, aside from the home and office, where you can be alone. This can be the local coffee place, a nearby park, a spot in the library, wherever that allows you to leave distractions behind. It can even be while you are on the move. Charles Dickens used to walk over 12 miles (almost 20 km) on average every day through the lush Kent countryside or the bustling streets of Victorian London. These walks gave his mind a freedom to wander, with many details and experiences ending up as fodder for his numerous novels.[105]

The American poet, playwright and activist Maya Angelou was also steadfast in carving out her Solitude Moment every day. From 7 am to 2 pm, she would be in a spartan hotel or motel room near home, "a tiny, mean room with just a bed, and sometimes if I can find it, a face basin", her third space where she could be deep in silence and thought.[106] Said Angelou of her daily ritual: "If the work is going badly, I stay until 12:30. If it's going well, I'll stay as long as it's going well. It's lonely, and it's marvellous."[107]

It does not matter what form our solitude comes in, or how long it takes. Some may take long walks, others may book a room at the Ritz-Carlton, while still others may pitch a tent at the beach. What matters is purposefully making time for our Solitude Moments.

> **"Without great solitude,
> no serious work is possible."**
> Pablo Picasso

Sleep as an Elixir

The second ritual for Next Practice is vital to any human being's wellbeing, more so for those of us who want to do some serious creative work.

Sleep. We are simply not getting enough of it.

The society we live in places a high value on productivity. We squeeze every last second out of our 24 hours to get more done, with sleep becoming the sacrificial lamb. We wear sleeplessness like a badge of honour, and fake-brag about our ability to keep going despite the bare minimum number of hours we sleep each night. Sleep? Who has time to sleep? It is lazy, unproductive and frankly, a waste of precious time.

Yet the science behind getting enough rest has shown how important sleep is. It is neither an indulgence nor a nice-to-have; it has been proven without a doubt to be crucial for our brains and bodies to function at an optimal level.

Tesla founder Elon Musk once said his mental sharpness dropped if he did not get a certain amount of sleep the night before, while Huffington Post's Arianna Huffington collapsed from sleep deprivation due to 18-hour work days that resulted in sleeping way less than eight hours a night.

On the flip side, Facebook's Mark Zuckerberg is keenly aware of the need for sleep. He wakes up at 8 am, sometimes later if he stayed up late chatting with programmers. He knows that one can still get work done without needing to put in punishing hours and forgoing sleep.[108]

Creators and inventors have also been known to tap into the creative space that sleep gives. Paul McCartney of The Beatles said that he came up with the melody for 'Yesterday' in a dream, while Elias Howe, the American inventor, had a nightmare – and from it came the inspiration for the sewing machine.

"Dreams are just thinking in a different biochemical state," said Harvard University psychologist Deirdre Barrett, author of *The Committee of Sleep*. "In the sleep state, the brain thinks much more visually and intuitively."[109]

Sleep is when we make mental connections and synthesise information in new ways. When we are sleeping, the brain has a sophisticated way of identifying important information from the day's bombardment of stimuli and facts, sorting through them and discarding the unimportant. This happens during slow-wave sleep, the kind that is most apparent in the early parts of our sleep cycle.[110]

In a 2004 study from the University of Lubeck in Germany, researchers had subjects complete math problems that relied on algorithms. What they did not know was that hidden deep within the formulas was an arithmetical shortcut. A quarter of the subjects discovered it on their own, but that number jumped to 59% when they were given a chance to get eight hours of sleep and then come back to solve it again.

Said cognitive neuroscientist Howard Nusbaum of the University of Chicago: "If you have an idea about a simpler solution and it's been working itself out in your head, you still tend to use the familiar one. When you sleep, the better answer has a chance to emerge."[111]

Not only does sleep help us come up with better solutions, it has been shown in particular to be a font of imagination. Penny Lewis and her colleagues from Cardiff University discovered that REM (Rapid Eye Movement) and non-REM sleep worked together to find unrecognised links between what we already know, while finding unusual solutions to problems that have been bothering us for a while.

"Suppose you're working on a problem and you're stuck," said Lewis. "[In REM sleep] the neocortex will replay abstracted, simplified elements [of the problem], but also other things that are randomly activated. It'll then strengthen the commonalities between those things. When you wake up the next day, that slight strengthening might allow you to see what you were working on in a slightly different way."

Lewis found that non-REM sleep extracts concepts from what is going on in our minds, while REM sleep connects them. And more importantly, both phases of sleep build on each other. Through the night, two parts of the brain in particular – the hippocampus and neocortex – repeatedly sync up and decouple, and the sequence of abstraction and connection repeats itself.

"The obvious implication is that if you're working on a difficult problem, allow yourself enough nights of sleep," she shared. "Particularly if you're trying to work on something that requires thinking outside the box, maybe don't do it in too much of a rush."[112]

What if despite the data for sleeping enough at night, some of us still cannot manage to chalk up enough hours at night? Taking naps is the next best option.

The Italians have long been known for their mid-day naps (perhaps these helped the creative genius of artists such as Michelangelo, Botticelli and Caravaggio), as do people in some other Mediterranean and Latin American countries. Mainland Chinese also have a culture of taking an afternoon nap, with a saying, *Zhongwu bu shui xiawu bengkui,* 'at noon if you don't sleep, in the afternoon you'll crash'.

Winston Churchill, even at the height of World War II, would retire to his private room after lunch, undress, sleep for an hour or two, then head back to 10 Downing Street for a shower and a change of clothes, before continuing with work. Churchill's valet, Frank Sawyers, later recalled: "It was one of the inflexible rules of Mr. Churchill's daily routine that he should not miss this rest."[113]

Why are naps so good for us? Certainly, as with nighttime sleep, it increases our alertness and reduces fatigue. Regular catnaps improve our memory over time, and it does not even matter the length of time taken. While more is usually better than less, countless studies have shown that whether it is a 5-minute nap or an hour-long nap, the benefits to memory retention, cognitive performance, imagination, emotion regulation, among other things, are immense.

Sara Mednick, psychologist at the University of California, Riverside, found in studies conducted with her team that people performed just as well on tests after a 60- to 90-minute nap as compared to after a full night's sleep. "What's amazing is that in a 90-minute nap, you can get the same [learning] benefits as an eight-hour sleep period," she noted.[114]

Other studies conducted by Matthew Walker of the University of California, Berkeley, and his team further found that sleep functions like a cleansing tool of sorts, helping to clear the brain's short-term memory in the hippocampus so that there is room for new facts to be remembered and moved to the prefrontal cortex region.

Said Walker: "It's as though the email inbox in your hippocampus is full and, until you sleep and clear out those fact emails, you're not going to receive any more mail. It's just going to bounce until you sleep and move it to another folder."[115]

Next Practice depends on cognitive agility. When we fail to listen to our body's need for sleep, and keep pushing it to accomplish more and at a faster pace, we are only setting ourselves up for failure. Our brains need the rest so that we can tap into its inherent genius. Sleep and taking naps are powerful tools we can easily use. If even during Britain's darkest hours of war, Churchill found time to nap, surely we can learn from it the next time we find ourselves stuck with a difficult problem or needing to move to the next level of innovation.

Moving to Our Cognitive Best

Like sleep, exercise has long been touted for its numerous benefits to our health on many levels, be it physical, mental or emotional. We know it keeps our heart working well, our risk for various illnesses low and releases a variety of hormones that make us feel happy.

However, not as much attention has been given to the impact of exercise on our brains. Exercise can protect our brain health. Exercise can make us smarter.

In the late 1990s, Henriette van Pragg and a team of researchers at the Salk Institute for Biological Studies in La Jolla, California, found that mice who lived in cages with toys and running wheels developed more new neurons in a process called neurogenesis. Even more fascinating were the results that showed that the mice that ran, compared to those that swam or tried to figure out a maze, had double the number of new neurons than the rest of the mice. These same mice also had greater reorganisation of their synaptic connections, suggesting that their running influenced the plasticity of their brains.[116]

Exercise can also make us more creative, which provides an extra boost in our search for the Next Practice.

"There's a lot of science behind this," said Scott McGinnis, an instructor in neurology at Harvard Medical School. Exercise stimulates physiological changes such as reductions in insulin resistance and inflammation while encouraging production of chemicals that affect the growth of new blood vessels in the brain, and eventually the abundance, survival and overall health of new brain cells. Moreover, people who exercise have larger hippocampuses, which is associated with memory and learning, compared to people who do not exercise.[117]

"Even more exciting is the finding that engaging in a program of regular exercise of moderate intensity over six months or a year is associated with an increase in the volume of selected brain regions," said McGinnis.

Researchers at Austria's University of Graz further found clear links between regular exercise and the human imagination. Lifestyles got healthier, moods lifted, and innovative thinking increased. Movement in general was clearly shown to fuel original and abstract thoughts.[118]

While it is clear that exercise does wonderful things for the brain, some of us may wonder: Does it matter what kind of exercise works best for boosting brain power?

Researchers at the University of Arizona suspected that running could be more intellectually demanding than a number of other exercises, and with the help of 22 young men – runners and those who did not exercise – found that the runners' brains compared to the sedentary men showed connections in the areas needed for higher-level thought. There was greater connectivity in the parts that aid in working memory, multi-tasking, attention and decision making, among other things. Remarkably, there was also less activity in the part of the brain that indicates a lack of focus.[119]

The suggestion from this and other studies does indicate that running is ideal as a ritual for awakening our System 2. As bestselling Japanese author Haruki Murakami once quipped: "Most of what I know about writing fiction, I learned by running every day."

Understandably, exercise (and particularly, running) is not everyone's cup of tea. What I do recommend to my students and clients is to keep exercise for the times when they need an extra boost in their performance, such as while dealing with an important assignment or project. It is like a vitamin they can consume when needed.

Solitude, sleep and exercise are three key rituals anyone interested in finding the Next Practice can consider practising regularly in their lives. Use them, enjoy the vast benefits that these three simple activities afford – and watch the magic happen.

Growing Our Myelin

There is one other secret ingredient to the mix that as Next Practice practitioners, we should be paying very close attention to.

Remember the thick white insulation sheaths called myelin that we explored in Chapter 4? Covering the neural pathways in our brains, this substance has been proven to help our brains work faster and better. Unsurprisingly, it is a key component in our brain-boosting training to activate our System 2.

Like the three rituals mentioned above, there are three easy practices that can improve the myelination process.

Firstly, thinking with our hands.

Can you remember what it was like to write with pencil and paper when you were a child? The deliberate control of the pencil to form the necessary strokes, the intentional making of prints on paper, the concentration and focus needed to express in writing what was going on in your head.

To an onlooker, what they would have observed was a child bent over a piece of paper, engaged in the simple act of writing. But tucked away in the brain, there would have been a buzzing hive of activity happening at the same time across both hemispheres, and activating the most unusual regions of the brain – the 'reading circuit' of linked regions.

These regions come alive when we are reading, and these are the exact regions that have been found to be activated during handwriting and not typing. Cursive writing in particular is what trains the brain to integrate multiple functions in sensation, movement control and thinking, because it is more demanding compared to normal handwriting. Writing basically improves our reading, and hence, knowledge acquisition. Furthermore, other research notes the hand's relationship with the brain when it comes to composing thoughts and ideas. In a study of children in Grades 2, 4 and 6, Virginia Berninger of the University of Washington found that those who wrote essays by hand instead of a keyboard wrote more words, wrote faster and expressed more ideas.[120]

In another study, which looked at Japanese university students and recent graduates from the University of Tokyo, Japan, researchers found that writing on physical paper rather than digital tools such as tablets and smartphones led to more brain activity, particularly in remembering the information an hour later. Subjects showed more brain activity in areas associated with language and imaginary visualisation, and in the hippocampus, the area linked to memory and navigation. The analog act of putting pencil to paper contained richer spatial details that could later be recalled and navigated in the mind's eye.[121]

The unique complex spatial and tactile information that physical writing brings forms our impressions and memory in a different way. Put simply, the sensory input from what we can see, feel and even experience is what changes our brains to retain and make sense of the information before us. Digital mediums that require us to tap, swipe and click do not come close in activating the hand-eye coordination nor the brain connections needed to boost our brain's agility.

Explained Professor Kuniyoshi L. Sakai, neuroscientist and author of the study: "Digital tools have uniform scrolling up and down and standardised arrangement of text and picture size, like on a webpage. But if you remember a physical textbook printed on paper, you can close your eyes and visualise the photo one-third of the way down on the left-side page, as well as the notes you added in the bottom margin."

Writing by hand engages our senses and motor neurons a lot more effectively than via the digital method. Nerve cells across more parts of the brain come alive, make more extensive connections and help us think faster and remember better.[122]

The researchers further encourage the use of paper for creative endeavours as well. "It is reasonable that one's creativity will likely become more fruitful if prior knowledge is stored with stronger learning and more precisely retrieved from memory. For art, composing music, or other creative works, I would emphasize the use of paper instead of digital methods," said Sakai.

The profession of architecture is a good example of how using the hands helps mediate between thought and reality. Hand drawing is essential to the process of design, allowing architects to quickly explore ideas and convey what is on their minds. Putting pen to paper is a creative process, allowing for keen observation, invention and problem solving. An idea evolves from the exploration stage through to the final sketch.

> **"Our hands, as Martin Heidegger proposes, are organs for thinking. When they are not working in order to know or learn, they are thinking. Drawing, building models, sketching... is a matter of 'doing' that turns into a way of 'thinking' where hands and ideas are joined together."**
>
> María Isabel Alba Dorado

The second practice to improve our brains' myelination is through cultivating optimum focus.

When we are able to put 100% of our concentration on a single problem or issue, that is when we are fully focused and present. We think we are 100% focused when we are working on something. Yet the data points to a rather different scenario.[123] A study by Microsoft Corporation noted that people generally lose concentration after eight seconds – the result of an increasingly digitalised lifestyle on our brains. The report states that heavy multi-screeners find it difficult to filter out irrelevant stimuli – they're more easily distracted by multiple streams of media.[124]

Our highly digitalised world has made it difficult for us to focus. The ability to focus, something taken for granted just a few decades ago, has now become a rare commodity. Reports show that on average, adults are no longer able to focus on a task for longer than 20 minutes at a time.[125] We consume so much of our content through online sources – in fact, online content consumption doubled in 2020 alone.[126] Yet, even though we are consuming more, we are also going through content in shorter form, be it tweets, Instagram posts or news articles. Our brains are being trained to focus long enough to pick up information quickly and in bite-sized pieces, and then to move on. However, the trade-off is our ability to go through long-form text, chew on more complex concepts that can only be expounded upon with the luxury of more words.

Digital screens have further stripped our experience of reading on paper. Despite attempts to recreate the tactile and sensorial experience of paper, digital screens fall far below the mark. Furthermore, the navigational difficulties of reading on a screen have been shown in numerous studies to impair our reading comprehension, and make it harder for us to remember what we have read when we are done. It also drains our mental capacity to focus due to the extra cognitive resources needed to make sense of the flat topography versus the more clearly defined topography of text on paper, the mental map of the text, so to speak.[127]

As if that is not bad enough, we also have to deal with the issue of *attention residue*. Sophie Leroy of the University of Washington has spent two decades studying the brain and how it deals with switching focus. While society extols the benefits of multi-tasking as a modern tool of productivity, what her research shows is that the brain generally finds it difficult to switch between tasks, especially complex and cognitively demanding ones.

Explained Leroy: "In particular, my research reveals that, as we switch between tasks (for example from a Task A to a Task B), part of our attention often stays with the prior task (Task A) instead of fully transferring to the next one (Task B). This is what I call Attention Residue, when part of our attention is focused on another task instead of being fully devoted to the current task that needs to be performed."

Whenever we get interrupted, have unfinished and pending work, attention residue easily occurs. Our minds keep them active on the mental backburner, instead of letting them go so we can focus on the task at hand instead. As a result of thinking about Task A, we have less cognitive resources available for Task B, which will likely lead to our performance on Task B suffering.[128]

It is easy to get entrapped by attention residue. We are constantly thinking about other tasks, always trying to fit more in by juggling a few things at once. Our lives are governed by never-ending to-do lists, many of which are filled with just basic daily administrative tasks, such as making appointments, sending emails, filling up application forms and the like. They can end up hanging around on our lists, undone, for weeks or even months, and occupy precious cognitive space.

Experiments published in 2001 by Joshua Rubinstein, Jeffrey Evans and David Meyer further reveal that in our brain's executive control process while working on something, there are two distinct stages – goal shifting ("I want to do this now instead of that") and rule activation ("I'm turning off the rules for *that* and turning on the rules for *this*"). These allow us to switch between tasks, but there is a switching cost to it, which while small, can add up if there are repeated and multiple switches. This is the downside to multi-tasking. These switching costs not only impact our productivity, but can also have serious repercussions, such as when a driver is switching between looking at his car navigation system and looking back at the road. That tenth of a second can potentially be all it takes for an accident to happen.[129]

A trick to navigating around attention residue is to use a tool that a group of Australian universities devised for their students, known as Get Your Life In Order (GYLIO) practices. This entails bundling tasks into a set amount of time, be it a morning, a day or a week, and simply focusing on clearing those tasks, one task at a time.

"If you have attention residue, you are basically operating with part of your cognitive resources being busy, and that can have a wide range of impacts – you might not be as efficient in your work, you might not be as good a listener, you may get overwhelmed more easily, you might make errors, or struggle with decisions and your ability to process information," noted Leroy.[130]

Our constant busyness clocking in long hours to manage the innumerable things we have on our plates is slowly tearing away at our ability to focus. The neural pathways of our brains do not have sufficient time to strengthen through sustained practice, before our attention is called away to some other task to work on, and myelination is interrupted.

Pang in his book *Rest* expounds on how creative people seem to have similar habits of work and rest. "Rather than working super long hours, they maximised the amount of depth of focus time they had per day, and really protected that and organised their day so they could put in about 4 or 4½ hours of really intensive deep work." That sweet spot of 4 plus hours varies from person to person, depending on individual differences in attentional networks and circadian rhythms, but there is generally a finite amount of time a day that our brain can really focus before it needs a timeout, before a lack of motivation sets in, mistakes increase and we get more distracted.

Borna Bonakdarpour, a neurology professor at Northwestern University shared that the main reasons for our limited ability to focus are cognitive overload and energy use. "When you increase the metabolism of the brain, it comes with by-products that need to be cleared out and cleaned. The brain needs to rest." According to Bonakdarpour, research shows that for every two hours of focused work, "you need about 20 to 30 minutes to break". If we combine this with solitude or sleep, it allows our DMN to kick into action, with our creative and subconscious mind continuing its work of coming up with insights.[131]

Focus is the most important part of our training in Next Practice. It is the execution part of Next Practice, when all that we have done to train our minds up till this point manifests in a tangible way. Without focus, even the best ideas cannot come forth to see the light.

Therefore, it is crucial that we regain our critical ability for optimum focus. One of the simplest ways is to start by reading a book. Yes, find a physical book, preferably a well-written classic. Read it slowly. Resist the urge to switch to something else, even if that moment comes in right at the start. Stay with the book, and aim for two pages. The next day, go with three pages. Incrementally, extend the number of pages you can stay focused on while reading until you find that you are able to sustain your focus for a much longer time than when you first began this practice.

The final practice I recommend to help build our brain's myelin for Next Practice is probably one that most of us would resist.

Put our smartphones away. Not simply out of our hands and on the table, but 'in the next room and out of sight' kind of away. Because our smartphones are making us dumber.

The research is compelling. Our smartphones are immensely distracting. When we use them, usually while multi-tasking (hello attention residue!), our attention, cognitive performance and productivity, among other things, plunge. For tasks that require greater attentional and cognitive demands, the use of smartphones cause even more pronounced hindrances.[132]

What is noteworthy is that the mere presence of a smartphone results in cognitive costs. In studies done with nearly 800 students, conducted by Adrian Ward of the University of Texas, Kristen Duke of the University of Toronto, Ayelet Gneezy from the University of San Diego, and Maarten Bos of Carnegie Mellon University[133], subjects were asked to focus on solving some problems. They were divided into three groups: one group kept their smartphones face down on the table, another kept them in a pocket or bag, while the last put them in a separate room. All smartphones had their sound alerts and notifications turned off. Nearly all the students said they were not distracted by their phones, but the results showed otherwise. The ones who performed best were those who had left their smartphones in another room. And you guessed it — the ones who fared the worst were those who had their phones on the table.[134]

Just having a smartphone within reach reduced available

cognitive resources. Even when people could resist the temptation of checking their smartphones, or when the devices were switched off, the impairment on cognitive capacity was the same, which was on par with the effects of sleep deprivation.

Why the lure of smartphones? Like the sirens in Greek mythology who seduce with their sweet voices, our smartphones figuratively call out our names, pulling on our attention. Research in cognitive psychology shows that human beings automatically pay attention to what is habitually relevant to us, much like a parent automatically turns towards a baby's cry. Our smartphones are our hubs of connection to the rest of the world; they keep us relevant to the world and vice versa. The more connected we feel to our smartphones, the harder it is to stay away from them. The mere effort put into trying to resist this gravitational pull uses our cognitive resources, and ironically, undermines our cognitive performance and leaves less for us to learn, reason and develop creatives ideas.

The solution is simply to put our smartphones away in a separate room when we do not directly need to use them, especially when we are in the process of ideation and imagination. We must free up as much mental capacity as we can to allow our neural pathways space to myelinate, strengthen, and basically do what they are meant to do.[135]

Paralysis by Fear of the Unknown

As a human race, we live our lives punctuated by a variety of fears.

There is the classic fear of public speaking (some say it is the Number 1 fear that most people have). Then there are others like the fear of snakes, of flying in planes, of germs and of many other things on a seemingly inexhaustible list.

However, if we are to peer deeper into each of these fears, and identify the *fundamental* fear, there is one common thread that runs through it all. It is the Fear of the Unknown (FOTU).[136]

We do not know how people will judge us when we are up on stage, nor whether the snake will hurt us, or whether the plane we are in will crash. Our fear of the unknown behind each of these situations is what causes the transference of that fear to another object, be it an audience, a snake or a plane.

Studies have also shown that we are unable to clearly see the difference between the stress that comes from a real danger and a perceived one. In our mind's eye, they are indistinguishable. Our brains activate the release of chemicals and hormones like cortisol to increase blood pressure, speed up our heart rate and stimulate our muscles. The amygdala, our instinctive brain, detects FOTU as real and takes over, and we default to fight, flight, freeze or avoid.[137] While we become more responsive, it is not possible to sustain that level of alertness for long, and soon, our bodies break down.

153

Insomnia, anxiety, paralysis of action, wrong choices and an inability to move forward with confidence are among some of the manifestations of those fears.[138]

Our System 1 operates from an illusion of understanding. Its automatic response to the world around us stems from an illusion of understanding; it believes what it knows is true and right. However, when FOTU comes into play – and it frequently does – they both combine to immediately reject new ideas that may inadvertently slip in from System 2. Human beings do not like change. We are much more comfortable being with the familiar and what is known. What this essentially means is while we can try to strengthen our System 2 by engaging in rituals of solitude, sleep and exercise, and practising methods to increase myelination of our brains, if we are unable to break free of our System 1, our efforts will come to naught. It is thus imperative that we minimise our FOTU so that we can escape System 1, and advance towards the Next Practice.

Now, as with much of what we have explored in this book so far, cognitive changes are notoriously difficult to make. I propose a three-step systematic method to reduce our FOTU.

The first is through exposure therapy, a psychological treatment that was developed to help people confront their fears. It is a simple concept – if someone has a fear, the psychologist will create a safe space for that person to be exposed to the fear in varying types and paces, such as virtual reality exposure and systematic desensitisation, to help reduce fear and avoid avoidance. In the case of Next Practice, if for example we fear artificial intelligence (AI) because we hardly know anything about it and worry that it brings with it insurmountable challenges, we could consider *in vivo* exposure by taking a short AI course, and imaginal exposure by imagining ourselves explaining AI to people we know. These steps would help us get more comfortable with learning and talking about AI, and eventually reduce our FOTU surrounding AI.[139]

The second method to consider is to find and use a safety signal. A 2019 report from researchers at Yale University and Weill Cornell Medicine showed that using a safety signal can help combat the anxiety that triggers bring on.

"A safety signal could be a musical piece, a person, or even an item like a stuffed animal that represents the absence of threat," said Paola Odriozola, co-first author of the report. In the studies done, both human and mice subjects that had access to safety signals showed a different neural network activated from those in exposure therapy. So doing things such as making notes before a presentation, praying before an exam or having coffee before an interview are all safety signals that we can use to soothe our nerves before our amygdala hijacks our minds.

The final method is finding and reframing our cognitive distortions. We all have them from time to time, and therefore need to be able to identify the key cognitive distortions that impact us, before finding the solution for eradicating it. Are we being plagued, for example, by polarised thinking, overgeneralisation, catastrophising or emotional reasoning? Sometimes the solution to that can be as simple as reframing or doing cognitive therapy such as exposure therapy.[140]

So go on, give this three-pronged method a try to see whether it helps you minimise your FOTU towards new (sometimes impossible) ideas.

Giving Voice to Ideas

Brainstorming is usually the preferred modus operandi in organisations when there is a need to spark group imagination. Teams are formed to generate as many ideas as possible, with unusual and original ideas in high demand, before going through a refinement and elaboration exercise.

The assumption is that having more people will lead to the stimulation of more ideas, and the more ideas there are, the higher quality the ideas that will eventually emerge. However, numerous studies have debunked the efficacy of this tool and show instead that brainstorming can harm creative endeavours. Brainstorming in large teams can actually backfire and harm productivity, while unwittingly resulting in groupthink.[141]

Allow me to propose a new way of thinking that brings together all we have learnt in the last few chapters. Instead of brainstorming, I suggest using what I call the Idea Voice.

This is a method that enables every individual in a team to have an equal chance at forming, and sharing, their ideas.

Just as in the practice of Next Practice, the key principles we are to observe in Idea Voice include:

1. Make mistakes
2. Individual Solitude Moment should precede team discussion
3. Psychological safety needs to be established before collaboration begins
4. Thinking with your hands (sketching, writing, etc.) is the way to go
5. Quantity of ideas trumps quality of ideas

Once we have spent enough time alone to first think through a problem, and have come up with our list of ideas, we move on to the actual steps of brainstorming in a manner that does not artificially give any idea extra weight or traction. Here are some ways we can accomplish that, coupled with the rules to abide by for each way of presentation.

A. The Voice Style Presentation

The presenter presents his/her idea by standing *behind* the audience (the audience can only see the PowerPoint and hear the voice of the presenter). Members take turns to be the presenter.

B. Idea Poster

The presenter summarises his or her idea using words, charts, drawings, etc., in one flipchart or one sheet of paper. The idea poster is then put up on the wall for everyone in the audience to see and vote for the best idea.

C. Sliding Vote from 1 to 4

Each team member will give a vote to each idea based on two categories – novelty and implementation. Novelty captures the newness of ideas while implementation looks at capturing the ease with which an idea can be executed, on a scale of 1 (not new) to 4 (never seen before).

The idea with the highest combined score will be declared the winning idea.

Idea Voice rules:

1. No judgement until voting. No comments, no body gestures, no facial expressions. Keep a poker face, whether you like or dislike the idea.
2. After the winner is announced, try to build on the ideas of others to make them better. Think 'and' rather than 'but'.
3. In any discussion, make sure to have only one conversation at a time. Every team member must have equal airtime to speak. No member should dominate the conversation.

4. Be visual. Draw your ideas, as opposed to just writing them down. Stick figures and simple sketches can say more than words.
5. Aim for idea quantity not quality. The best way to find onc good idca is to comc up with lots of idcas.

So there we have it. A blueprint for what we can do to prepare our minds' escape from System 1 into a robust, healthy and inherently creative System 2. When we can implement all the recommendations in this chapter, the brilliance of Next Practice will finally be there for the picking.

> **"We are what we repeatedly do.**
> **Excellence, then, is not an act,**
> **but a habit."**
>
> Aristotle

PART III

Creating Next Practice

from the Outside In

CHAPTER EIGHT

Design Thinking and Working Backwards + Next Practice

Mazes are tricky things.

They start off deceptively simple, but once you are inside, they become giant brain-teasers, filled with twists and turns that bring you on multiple paths in an almost haphazard fashion. The aim of a maze is to get through it to the end, but in order to do that, we must get lost a few times, hit dead ends, retrace our steps and choose different paths, and get comfortable being totally confused, before we reach the exit. In fact, the word 'maze' dates from the 13th century and comes from the Middle English word *mæs*, which means delirium or delusion.

That, in a nutshell, describes the process of innovation. We start off armed with a sexy results-guaranteed framework, buoyed by optimism that we will find the next big thing. Using a variety of techniques and strategies, we labour painstakingly but end up with good enough, but not great, ideas. The Next Practice eludes us.

Perhaps the reason is we are still searching among the scarce fruits of a high burden of knowledge, and therefore come up short. There are way too few fruits for the picking.

Next Practice calls for avant-garde courses of action, ways of thinking. Relying on prior knowledge and competencies as benchmarks will just keep us within the safe zones of what we know, instead of bravely venturing into the Gardens of Eden of innovation (Chapter 1).

What if instead of just making Incremental Innovations, we scan the landscape for ideas with a low burden of knowledge, to discover the Architectural Innovations and Disruptive Innovations, or even better, the Radical Innovations?

What if we relook the start point of this maze of innovation-seeking, and apply what we have learnt about Next Practice to supercharge our search?

I will use four examples of innovation processes to show you how Next Practice principles can be applied to supercharge them to the next level. In this chapter, we will examine how the processes of Design Thinking and Working Backwards – which by nature are exercises in the positivity of making mistakes (Chapter 6) – can liberate us from the discomfort of failure, and bring us closer to discovering our Next Practice.

Design Thinking: The Problem Deciphering Engine

Design thinking is a people-centric method for solving problems, especially wicked problems, or those that are ill-defined. In a five-step non-linear iterative method, design teams Empathise, Define, Ideate, Prototype and Test innovative solutions that can meet their consumer's needs.[142] These five steps can happen in any order, be repeated, and even worked backwards, all with the aim of arriving at a crystal-clear understanding of the problem at hand.

The consumer is king, and his needs are placed above all else. Assumptions must be challenged and problems redefined right from the beginning. Why? Because the wrong questions will lead to the wrong answers. Surveys of 106 public and private sector C-suite executives from 17 countries showed that 85% agreed that their organisations were bad at problem diagnosis, and 87% agreed that this carried significant costs.[143]

Therefore, it is imperative to start with the right questions. Creative solutions nearly always come from alternative definitions of the problem and questions. How can we do this?

A key strategy for accomplishing this in Design Thinking is through the use of an Empathy Map, a collaborative visualisation of who the consumers are. A grid with four quadrants – Says, Does, Thinks, Feels – is first drawn. Using Post-Its, the working team will capture their observations, one observation per Post-It, and place them in the appropriate quadrants. What is the profile of this consumer? What does he need and how does it feel being in his shoes? What insights and observations can be made about his thoughts, feelings and actions as he interfaces with the context of the problem? Are there any surprises?[144] This mapping exercise and the subsequent conversations that follow draw out patterns and unknowns in order to do a really deep dive into the person of the consumer, and finally arriving at a clear understanding of what the business needs to do for that consumer.

The Empathy Map is the linchpin of Design Thinking. Empathy is the ability to understand the experiences and reality of another, by sharing their perspectives without getting lost in them. A significant amount of research has shown that the practice of mindfulness can promote an increase in empathy. The regions of the brain needed to improve empathy – the prefrontal cortex, anterior cingulate cortex, anterior insula – are developed with the practice of paying attention to one's inner experience, such as in mindfulness practices. Furthermore, mindfulness increases understanding of one's emotions and the present moment, in a non-judging manner, all of which contribute to empathy for another.[145]

In Chapter 5, we explored the power of meditation in activating the Default Mode Network (DMN), the part of the brain that is the seat of imagination. If we want to grow our empathy muscle so that we can create more accurate Empathy Maps in Design Thinking, it stands to reason that starting a Next Practice ritual of regular mindfulness exercises will help. It has been shown that the more we engage in analytical thinking, the less we can tune in to empathic thinking. The brain areas that allow us to understand other people's experience get turned off; there is a neurological divide.[146]

I often advise my clients and students to consider doing a walking meditation every day, for a few days, before doing the Empathy map. The DMN will naturally kick in to enhance their discovery of the right questions to ask and find the answers to.

The process of writing ideas down while doing the Empathy Map further enhances our imagination, allowing us to think better and faster, all with the simple act of thinking with our hands (Chapter 7) through writing on Post-Its.

Lastly, the power of Design Thinking relies on its iterative non-linear nature, or looping. Design asks 'what could be', not 'what is', in order to deal with problems that have no predetermined answers. The loop therefore is a beautiful fluid cycle of observing, reflecting and making. We understand the present, envision the future. We build on successes, make mistakes. In Chapter 6, we saw how failure and success are intertwined for Next Practice – one has to come with the other. Mistakes are par for the course. They are the essence of looping in Design Thinking.[147]

NEXT PRACTICE TIPS

1. Use walking meditation, or a meditation practice of your preference, to activate your DMN.
2. Write on paper as often as you can, with the Empathy Map exercise or even in other segments of your Design Thinking process, to tap into your imagination and increase brain connections for quicker and better responses.
3. Make sure to relax. Go and star-gaze. Watch a funny movie. Or pick something that you prefer from the list of Relaxation Techniques in the Appendix.
4. Get comfortable with making mistakes, trying out the most unusual audacious ideas. Diverge with no holds barred, and converge again. Keep the momentum going until the Next Practice solutions are found.

METHOD BEST FOR?

Design Thinking is a process most suitable for teams that want to look at creating Architectural Innovations and Disruptive Innovations. IBM and Airbnb are classic success stories of this approach.

Working Backwards: Back-to-the-Future Vehicle

Recall the concept of the maze that we started this chapter with.

Most of us tackle mazes by beginning at the start point, to get to the end. What if we reversed that order?

Famed German mathematician Carl Gustav Jacob Jacobi followed a simple strategy in his work – *man muss immer umkehren* (loosely translated to mean 'invert, always invert').[148] Many hard problems are best solved when addressed backwards, because it forces us to consider different perspectives that are blind spots when we approach them from the usual forward-looking standpoint. We see things in a different light when we use the inversion principle.

For example, instead of thinking about how to be rich, we should ask what can be done to avoid being poor. Possible answers could range from curbing overspending, having a budget to stick to, staying away from habits that require a lot of money. These would cut expenses, and leave us more cash-rich. Over time, can we become rich? Of course. Did we have to work extra jobs to gain that extra money? Not at all.

Or if I want my employees to be more productive, ask instead what makes them unproductive. Constant interruptions, unnecessary meetings, busy being busy, etc., all make one unproductive. By eliminating these, my staff should quite naturally have more time to think and increase their productivity.[149] It is sometimes easier to avoid what we do not want, than to get what we want. This is Via Negativa in action.

Jeff Bezos used this same inversion principle when deciding whether to leave a high-paying job to start Amazon. "When you are in the thick of things, you can get confused by small stuff," he recalled. "I knew when I was eighty that I would never, for example, think about why I walked away from my 1994 Wall Street bonus right in the middle of the year at the worst possible time… [but] I knew that I might sincerely regret not having participated in this thing called the Internet that I thought was going to be a revolutionising event. When I thought about it that way… it was incredibly easy to make the decision."[150]

So rather than ensuring his happiness, Bezos looked to reducing his regret. He inverted his perspective to gain clarity, and imagined his future. It is this same approach he used to grow Amazon to the massive e-commerce and technology conglomerate it is today, worth a staggering US$1.76 billion.

As with his personal eureka moment, Bezos realised the power of inversion in business. At Amazon, their "We are obsessed with our customers" culture means that the entire focus is on delivering what the customer wants and needs. That is the end point, and that is where Amazon begins from. With the end in mind.

Ian McAllister, Director of Amazon Day, shared that Amazon's approach towards product development starts from Working Backwards – "from the customer, rather than starting with an idea for a product and trying to bolt customers onto it".[151]

Working Backwards relies on imagination. In starting from the end, product managers must look to the future and consider how it could possibly look.

The process for new initiatives always begins with writing an internal press release about the finished product. "Internal press releases are centred around the customer problem, how current solutions (internal or external) fail, and how the new product will blow away existing solutions," said McAllister. "If the benefits listed don't sound very interesting or exciting to customers, then perhaps they're not (and shouldn't be built)." The press release is revised over and over again until the idea has been refined to satisfaction, before the product goes into development stage.

That is a lot of work for an idea that may never be developed. However, iterating on a press release at the start will cause a lot less pain than iterating on a product that will eventually not fly with customers. Product managers will also need to create a comprehensive list of FAQs to address both internal and external stakeholders; this list, too, goes through the same rigorous revisions.

> **"Done correctly, the Working Backwards process is a huge amount of work. But, it saves you even more work later. The Working Backwards process is not designed to be easy, it's designed to save huge amounts of work on the backend, and to make sure we're actually building the right thing."**
> Jeff Bezos[152]

At Amazon, they always asked: What were customers' needs? Did they have the necessary skills to build something that met those needs? If not, how could they build or acquire them? They never let their lack of ability deter them from achieving the end result of meeting customer needs.

Working Backwards was what enabled products and services such as the Amazon Web Service, Amazon 4-Star Store, and cloud computing, for which the technology did not even exist at that point. They simply built their capabilities from scratch. Nowhere was this clearer than with the story behind the Kindle.

Amazon knew that their customers wanted the e-book equivalent of an iTunes/iPod experience: an app paired with a mobile device that offered consumers any book they wanted, at a low price that they could buy, download and read in seconds. But in order to do that, they had to invent the hardware for it. Amazon was an e-commerce store, not a hardware store. They knew nothing about building hardware. Yet they pressed on. And the rest is history.[153]

I will leave you with a final story on the power of Working Backwards.

The Higgs Boson is known as the 'God particle', the fundamental particle associated with the Higgs field, a field that gives mass to other particles such as electrons and quarks. But like many other great scientific discoverers, scientist Peter Higgs imagined the existence of this particle almost five decades before it was finally discovered and proven by the Large

Hadron Collider in 2012.[154]

How was he able to imagine something so significant, without any indication of its existence? And how was it finally proven real?

We, too, can have our own Higgs Boson moments. With the power of Next Practice in Working Backwards.

NEXT PRACTICE TIPS

1. Use Via Negativa, or the inversion principle, to uncover our hidden beliefs about problems.
2. Tap on our Quiet Brains (Chapter 5) and R&T (Chapter 7) of Next Practice to get past our assumptions and illusion of understanding. Our DMN is the source of imagination that will enhance the discovery of our end in mind.
3. Make sure to relax. (Pick something from the list of Relaxation Techniques in the Appendix.)
4. Focus on using Working Backwards as a device to flesh out ideas, test out alternatives and be liberated from the norms of a forward-looking approach.

METHOD BEST FOR?

Working Backwards is a process most suitable for teams that want to look at creating Radical Innovations. Amazon is the poster child for using this method to great success.

CHAPTER NINE

Appreciative Inquiry and First Principles + Next Practice

The Hindu religion has a beautiful depiction of their one supreme god, Brahman, from which everything comes.

In that depiction stands the Trimurti ('three forms' in Sanskrit), the trinity of gods: Brahma the Creator, Vishnu the Preserver, and Shiva the Destroyer. While Brahma's work is considered done with the creation of the world, Vishnu and Shiva continue their work.

Much like the roles of Vishnu and Shiva, the natural order of the world dictates a necessary cyclical maintenance and destruction. Of growth and maintenance, coming apart and rebuilding, life and death.

So, too, does innovation follow this order. There is a start, a middle, an end – and a start again. On and on it goes.

Each has its brilliance. Yet sometimes, we forget this. We are enthusiastic about starts and in-betweens, but are resistant to endings. Ending something is hard to do; it is as if human beings are programmed for life, not death. No one likes destruction. So, we either push forward toward newness all the time, or preserve the status quo of assumptions, without ever stopping to ask: is it time to kill the current, in order to have a better start again? Are we even starting from the right point?

In this final chapter, we will examine two more processes, Appreciative Inquiry and First Principles, the exemplification of the preserver and destroyer in innovation respectively, and in light of the Next Practice approach, see how we can turbo-charge them.

Appreciative Inquiry: Building Strength from Within

> **"Ever since I was a teenager, I've tackled every big new problem the same way: by starting off with two questions... Who has dealt with this problem well? And what can we learn from them?"**
> Bill Gates[155]

This is a classic example of the Appreciative Inquiry framework in action.

For companies that use Appreciative Inquiry, much like how Gates approaches huge problems and even goals, the strategy is a strengths-based one that does not waste time reinventing the wheel. Instead, it looks for whatever is already working well at the moment. What are the existing assets in the business? Where have we done well? What knowledge do our people currently have, and how can our strengths help us face the challenges in future? How can we use what we know to build on our positive experiences for the road ahead?

Piql is a Norwegian company that converts data, be it written text or images, into binary code and data points for digital storage – on old school physical film! This technology uses photosensitive film called piqlFilm which is then stored in a film cartridge, or piqlBox, an offline, physical storage format that has a purported lifespan of 500 to 2,000 years, without degrading.[156] Film had been seen as a dying industry, a has-been that was good while it lasted but was no longer needed. Yet Piql took an Appreciative Inquiry approach to looking at the strengths of film, and breathed new life into a dead product that did not seem to have a future.

The Appreciative Inquiry method is one that changes our focus and conversations from deficit-oriented to asset-oriented. Instead of asking negative questions such as "What did these low-performing offices do wrong?", the method encourages questions like "What did these high-performing offices do right?". The reactions to the latter would be energising, uplifting, and take us on a different planning process, accessing different knowledge compared to if we took a problem-focused approach. It becomes much easier to design a future we find compelling, to tell stories and draw images of the positive future we see. The more people share and buy into that vision, the more probable that vision becomes.[157]

The challenge with Appreciate Inquiry is bringing people to the stage of envisioning that future. In the 4D stages of Appreciative Inquiry – Discovery, Dream, Design, Destiny – users often find it difficult to move beyond what they naturally focus on.

To discover and dream requires us to access our System 2 way of thinking. Human beings often instinctively look at what we do not have, our gaps and weaknesses (perhaps a legacy from our hunter-gatherer days when we had to constantly monitor what we did not have in order to survive!). Our preconceived, and often strongly entrenched, assumptions and biases through which we view the world can keep us from discovering what is good around us. To then take it a step further and dream about a positive future can be a giant leap for many. How does one dream about a future when the present is unclear?[158]

In order to unlock the true potential of Appreciative Inquiry, using Next Practice can help to bridge the gap. While in the Discovery stage, building our myelin (Chapter 4) to see connections where we never saw them before and allowing mistakes to be made (Chapter 6) in the process of searching can contribute to a more robust identification of strengths. And it goes without saying that the Dreaming stage would require a fair amount of engaging the Quiet Brain (Chapter 5) in order to see the future.

NEXT PRACTICE TIPS

1. Learn. Repeat. Take the opportunity to learn new things about your company, see things from different perspectives, chunk it down and assimilate the information.
2. Find time to rest for optimal myelin production – sleep!
3. Allow for many starts and stops, because this process of making mistakes will encourage the natural selection of the eventual best observations and ideas.
4. Be quiet, find places of solitude. Meditate. Even small pockets of solitude and meditation, done frequently, can create wonders for your DMN.

METHOD BEST FOR?

Appreciative Inquiry is a process most suitable for companies that do not want to take too much risk, and prefer to innovate from an environment of Incremental Innovations. They can move on from that point to create Architectural and Disruptive Innovations. Microsoft is one such example. However, using this process to achieve Architectural and

Disruptive Innovations is the hardest, as it tends to activate our illusion of understanding due to familiarity with what is already existing, and that means falling back into the trap of System 1 again.

First Principles: Alchemy of Revolution

Spend some time with a child and you will probably, rather quickly, have a conversation that goes something like this:

"Why are you drinking coffee?"

"Because I'm tired and I need something to help me stay awake."

"Why are you tired?"

"Because I was up late working last night."

"Why do you need to work till so late?"

"Because I need to work and earn enough money for my family."

"Why must you earn enough?"

Why, why, why. Drilling down to the fundamentals, the essence of understanding. Children have a knack for doing this, and in all their innocence, they do not realise the sophistication behind their line of questioning, known also as First Principles.

A First Principle is a foundational proposition or assumption that stands alone, and cannot be deduced from any other proposition or assumption. Aristotle defined First Principles as "the first basis from which a thing is known".[159] Socratic questioning is used to establish First Principles through strict and disciplined analysis, following a process that reveals hidden assumptions that can act as blinders.

Richard Feynman, the American theoretical physicist and Nobel Prize winner for his work in quantum mechanics, was a great proponent of First Principles. "The first principle is that you must not fool yourself – and you are the easiest person to fool."[160] He implicitly understood that First Principles thinking was a way to escape the illusion of validity and understanding of our System 1. For every problem he faced, he would break it down into its simplest form. Even his Feynman Learning Technique demonstrated this. The four-step method – pretend to teach a concept to a child (or rubber duck!), identify the gaps in explanation, organise and simplify, and finally present it to someone again – is First Principles in action, because it calls for drilling down a concept to its fundamentals.[161]

For example, if we were to apply First Principles (or even the Feynman Learning Technique) to creating a new dish from spaghetti bolognese, we would start by identifying the ingredients to explain what goes into the dish – tomatoes, onions, beef and pasta. What goes into the pasta? Flour, salt and oil. Can we do anything else with those ingredients, and form something new from it? Deconstructing the dish allows us to zoom in on the basic characteristics of each base ingredient, and from there, create a new dish we have not thought of before. This is Next Practice at play.

Reasoning by First Principles breaks down complicated

situations and problems into their most basic elements, looks at the assumptions we have about them, and then reconstructs them. Reasoning by analogy, on the other hand, is when we build on knowledge or solve problems based on prior assumptions and best practices that are generally held as acceptable by others.[162]

Guess which would result in a better result?

This approach of thinking is one touted by centibillionaire Elon Musk, founder of Tesla and SpaceX, as the secret to his success in innovation and imagination. Surely it must be working, for him to break through all that was thought impossible, creating a rocket that is able to land safely on Earth upon re-entry. This feat was not achieved by NASA even after more than six decades of prototyping and testing. SpaceX did it in less than 10 years, and with much less funding.[163]

The impossible is possible after all.

Musk points out that he always starts with what is true, not with his intuition, which can be unreliable since as humans, we often do not really know as much as we think we do. We are hindered very much by our assumptions.

"I think people's thinking process is too bound by convention or analogy to prior experiences. It's rare that people try to think of something on a first principles basis. They'll say, 'We'll do that because it's always been done that way.' Or they'll not do it because 'Well, nobody's ever done that, so it must not be good'. But that's just a ridiculous way to think.

"You have to build up the reasoning from the ground up – 'from the first principles' is the phrase that's used in physics. You look at the fundamentals and construct your reasoning from that, and then you see if you have a conclusion that works or doesn't work, and it may or may not be different from what people have done in the past."[164]

To him, most people get through life reasoning by analogy, "which essentially means copying what other people do with slight variations". What then happens is we will end up doing what others have always done, trapped in the way things have always been. Herd mentality sets in. Imagination shuts down. Innovation cannot happen.

Analogous thinking forces us to conform to the norms of society, make improvements upon what is in existence, and see things through the lens of others – who would also have seen things through the lens of others before them.

So much clutter.

First Principles cuts through all of that, and delivers a game-changing approach that helps us see what is truly possible. First Principles powers us away from small incremental steps towards giant leaps of possibilities that the analogous thinkers do not see.

Look at the example of Singapore Airlines, the iconic flag carrier airline of Singapore, leading innovator in the airline industry and voted Best International Airline 25 years in a row.[165] The Covid-19 pandemic hit it hard, as it did many other airlines around the world. However, true to its never-say-die form, Singapore Airlines pivoted quickly into trying out different services it had never offered before, in order to survive.

Its dexterity in industry-leading innovations over the years, as with its performance in the Covid era, is what has helped it survive and thrive over the last 50 years. As one of its vice-presidents put it: "Whatever we do, we are in search of excellence and are never willing to settle for what we have already achieved. We have to be able to tell ourselves that, 'I love this new thing that I've developed and we'll make sure that it's well implemented.'

"However, we also have to kill it with a better product in X number of months. It could be six months, it could be 12 months, it could be 20 months; but you have got to kill it because the lifestyles of our customers are continuously evolving."[166]

Now that is one classy implementation of First Principles – bold, never flinching from destroying the now, so as to rise from the ashes like the proverbial phoenix, to become stronger and better.

Imagine, what will be the outcome if we marry First Principles with Next Practice? Idea explosions. And not just any type of idea, but explosions of avant-garde ideas that are a class above the rest.

NEXT PRACTICE TIPS

1. Escape from our System 1 by finding ways to build our myelin production (Chapter 4).
2. Combine myelin production (Chapter 4), harnessing the Quiet Brain (Chapter 5) and allowing for mistakes (Chapter 6), and the synergy of this will unlock the full potential of First Principles.
   - The Quiet Brain will connect the right neurons to ask the right questions.
   - Myelin production will allow the right questions to surface.
   - Making as many mistakes as possible will allow us to find and select the best questions.
3. Rest. Solitude. Exercise. Repeat.

METHOD BEST FOR?

First Principles is a process most suitable for companies that are not content with just Incremental Innovations, or even Architectural and Disruptive Innovations. They want to create Radical Innovations. Think Tesla.

**"A man is what he thinks about
all day long."**
Ralph Waldo Emerson

Conclusion

I wanted to write this book because I believe in the quest for human ingenuity.

There is so much latent power in our minds, power that can move worlds and change realities. Not just for powerful nations and big corporations, but especially for the underdogs of the world — the countries in poverty, fledging start-ups and individuals struggling to feed their families and educate their young.

The luminance of our brains is waiting to illuminate, but there are many broken links that first need to be fixed. We cannot expect genius if we are not willing to first do our part in repairing those links — and then enhancing them some more. It takes two hands to clap.

Yes, the groundbreaking data and latest research, the rituals and training for Next Practice I speak of, all of that can create innumerable opportunities for those who are seeking the next big thing in business innovation and policy-making.

But the real game-changer that can come out of the learnings put forth in this book is the impact on the lives of many who, with simple and consistent changes in their daily routines, can have mega-breakthroughs that will forever change the trajectory of their future.

This is my hope. The hope that imagination brings. The hope that from my wish to change the world, I can first change the realities of that one man in the street.

Relaxation Techniques

Nature Techniques

Connecting with nature is soothing. Ever heard of the phrase "forest bathing"? It means simply being in nature. Being in nature can help clear your mind and is a great way to relax.

1. Hiking in the forest
2. Walking in the park
3. Driving a scenic road
4. Fishing in the lake
5. Camping in the mountains
6. Sleeping beneath the stars
7. Strolling on the beach
8. Watching the sunset/sunrise

Mind Techniques

What you think about determines your emotional reaction, so thinking of the worst-case scenarios, and criticising yourself over mistakes will all increase your stress level. Instead, allowing yourself to think of the best-case scenarios and accepting mistakes as learning opportunities will make you more relaxed.

9. Meditation
10. Mindfulness
11. Positive thinking
12. Reframing a problem
13. Daydreaming
14. Guided imagery

Martial Arts Techniques

Some martial arts are gentle in nature. These martial arts possess a gentle way of moving and stretching our body. Regular practitioners can benefit from not only a stronger body but also a more relaxed mind.

15. Taichi
16. Ba Duan Jin
17. Aikido
18. Kyudo
19. Capoeira

Having Fun Techniques

Fun activity is a source of *eustress*.[167] Eustress is a positive form of stress having a beneficial effect on emotional well-being. This is the good kind of stress that makes you feel alive and relaxed. We need regular eustress in our lives, and fun activities can provide that.

20. Watching a movie
21. Going to a theme park
22. Fine dining
23. Drinking a great wine
24. Window shopping
25. Attending a concert
26. Visiting a museum
27. Dancing in a club
28. Singing karaoke
29. Drinking dark chocolate
30. Playing computer games

Home Techniques

People often say "Home sweet home" when they arrive back in their own house, or "There's no place like home". Because of this unique nature, our home is the best place to relax.

31. Aromatherapy
32. Listening to music
33. Playing music
34. Baking
35. Gardening

36. Warm bathing
37. House cleaning
38. Practising yoga
39. Reading a book
40. Cuddling your pet
41. Sleeping in solitude

Spiritual Techniques

Spirituality has many benefits for stress relief – by helping you feel a sense of purpose, connect to the world, and lead a healthier life.

42. Devotion
43. Singing religious songs
44. Participating in religious service
45. Chanting a prayer

Art Techniques

Art is a great stress relief tool, even for those who don't consider themselves artistically inclined.[168] Creating art can take your mind off whatever is stressing you and make you relax during the process of art-making.

46. Painting a landscape
47. Sketching people
48. Writing a story/poem
49. Taking street photos
50. Making pottery
51. Colouring book

Breathing Techniques

Deep abdominal breathing encourages full oxygen exchange. The lowest part of the lungs thus gets a full share of oxygenated air. That can slow your heartbeat and make you feel relaxed.

52. Pranayama breathing
53. Zen breathing
54. Qigong breathing

Body Techniques

Massage therapy improves blood circulation. Improved circulation can enhance the delivery of oxygen and nutrients to muscle cells. As the result, your body enters relaxation response – a state in which your heart rate slows, your blood pressure goes down, your production of stress hormones decreases, and your muscles relax.

55. Body massage
56. Foot reflexology
57. Face mask
58. Ear massage

Sport Techniques

Exercise causes your body to release endorphins, the chemicals in your brain that relieve stress and make you feel happy and relaxed.

59. Swimming
60. Running

61. Biking
62. Surfing
63. Sailing
64. Skating
65. Bowling
66. Diving
67. Canoeing

Office Techniques

Office is not the best place to relax. However, you can still feel relaxed if you can pull the right tricks.

68. Making coffee/tea
69. Eating at the pantry
70. Chatting with colleagues
71. Stretching at your desk
72. Tidying up your desk

Digital Detox Techniques

Digital detox refers to a period when a person refrains from using digital devices such as smartphones, laptops, and social media sites. Research has found[169] that heavy digital device use among young adults is linked to sleep problems, depressive symptoms, and increased stress levels. Reducing the usage of digital devices will alleviate the problems.

73. Stop multitasking
74. Social media fasting
75. Taking a break from your smartphone

Relationship Techniques

The Harvard Study of Adult Development,[170] one of the longest-running studies on happiness, found a strong association between happiness and close relationships – spouses, family, friends, and social circles. Personal connection creates mental and emotional stimulation, which are automatic mood boosters that will make you happy and relaxed.

76. Calling a friend
77. Cooking for your boyfriend/girlfriend
78. Buying a gift for your husband/wife
79. Visiting your parents
80. Drinking with buddies
81. Volunteering with colleagues

Notes

1. Kahneman, D. (2013). *Thinking, Fast and Slow*. Farrar, Straus and Giroux.
2. West, D. (2017, February 14). How to predict the lifespan of a company in one simple measure. *ITProPortal*. https://www.itproportal.com/features/how-to-predict-the-lifespan-of-a-company-in-one-simple-measure/
3. Viguerie, S.P., Calder, N., & Hindo, B. (2021, May). 2021 Corporate Longevity Forecast. *Innosight*. https://www.innosight.com/insight/creative-destruction/
4. Sheetz, M. (2017, August 24). Technology killing off corporate America: Average life span of companies under 20 years. *CNBC*. https://www.cnbc.com/2017/08/24/technology-killing-off-corporations-average-lifespan-of-company-under-20-years.html
5. Ruprecht, J. (2020, 17 September). VUCA & Agile Leadership: Part 2. *agilityIRL*. https://www.agilityirl.com/vuca-agile-leadership-part-2/
6. Hietala, J., Harju, J., & Kuosmanen, S. (2019, March 15). Why do we spend all that time searching for information at work? *Valamis Technology*. https://medium.com/@valamis_technology/why-do-we-spend-all-that-time-searching-for-information-at-work-114c729aed7
7. Bloom, N., Jones, C., Van Reenen, J., & Webb, M. (2017, December 20). Great Ideas Are Getting Harder to Find. *MIT Sloan Management Review*. https://sloanreview.mit.edu/article/great-ideas-are-getting-harder-to-find/
8. Wong, M. (2017, September 14). Stanford scholars say big ideas are getting harder to find. *Stanford News*. https://news.stanford.edu/2017/09/14/stanford-scholars-says-big-ideas-getting-harder-find/

9.   Laurence, P. (n.d.). The Burden of Knowledge. *Ask Magazine.* https://www.nasa.gov/pdf/575358main_43kn_burden_knowledge.pdf

10. CNBC. (2018, June 29). CNBC Transcript: Nadiem Makarim, Founder and CEO, GO-JEK. https://www.cnbc.com/2018/06/29/cnbctranscript-nadiem-makarim-founder-and-ceo-go-jek.html

11. Nurhadi. (2018, May 16). The Tale of Gojek and Bluebird Cooperation: Who's Winning? *Nurhadi.* https://medium.com/@haddn/the-tale-of-gojek-and-bluebird-partnership-whos-winning-4ad054ea7d90

12. Rosenzweig, P. M. (1993, July 8). Bill Gates and the Management of Microsoft. Harvard Business School Case No. 9-392-019. Gino, F., Ciechanover, A., & Huizinga, J. (2020, November 15). Culture Transformation at Microsoft: From 'Know it All' to 'Learn it All'. Harvard Business School Case No. 9-921-004.

13. Bohnhoff, T. (2020, February 14). Bringing AI to everyone. *Appanion.* https://medium.com/appanion/bringing-ai-to-everyone-c1b235a6af1

14. Microsoft. (2014, March 27). Satya Nadella: Mobile First, Cloud First Press Briefing. https://news.microsoft.com/2014/03/27/satya-nadella-mobile-first-cloud-first-press-briefing/

15. Kemp, S. (2021, July 22). Half a Billion Users Joined Social in the Last Year (And Other Facts). *Hootsuite.* https://blog.hootsuite.com/simon-kemp-social-media/. Moshin, M. (2021, April 5). 10 Social Media Statistics You Need to Know in 2021 (Infographic). *Oberlo.* https://sg.oberlo.com/blog/social-media-marketing-statistics

16. Farrell, A. M., Goh, J., & White, B. J. (2014, June 10). The Effect of Performance-Based Incentive Contracts on System 1 and System 2 Processing in Affective Decision Contexts: fMRI and Behavioral Evidence. SSRN. https://papers.ssrn.com/sol3/papers.cfm?abstract_id=2111452. Kahneman, D. (2013). *Thinking, Fast and Slow.* Farrar, Straus and Giroux

17. Bretas, R. V., Yamazaki, Y., & Iriki, A. (2020, December). Phase transitions of brain evolution that produced human language and beyond. *Neuroscience Research,* 161: 1–7. https://www.sciencedirect.com/science/article/pii/S0168010219304882

18. Swaminathan, N. (2008, April 29). Why Does The Brain Need So Much Power? *Scientific American.* https://www.scientificamerican.com/article/why-does-the-brain-need-s/

19. Edwards, S. (2016). *Sugar and the Brain.* Harvard Medical School. https://hms.harvard.edu/news-events/publications-archive/brain/sugar-brain

20. Bratsberg, B., & Rogeberg, O. (2018, June 26). Flynn effect and its reversal are both environmentally caused. *PNAS*. https://www.pnas.org/content/115/26/6674

21. Bratsberg, B., & Rogeberg, O. (2018, June 26). Flynn effect and its reversal are both environmentally caused. *PNAS*. https://www.pnas.org/content/115/26/6674

22. Cuddy, A. (2017, December 29). The IQ of Europeans is dropping due to technology, say researchers. *Euronews*. https://www.euronews.com/2017/12/29/the-iq-of-europeans-is-dropping-due-to-technology-say-researchers

23. Aldric, A. (2020, October 1). Average SAT Scores Over Time: 1972–2020. *PrepScholar*. https://blog.prepscholar.com/average-sat-scores-over-time

24. Lewis, P. (2017, October 6). 'Our minds can be hijacked': The tech insiders who fear a smartphone dystopia. *The Guardian*. https://www.theguardian.com/technology/2017/oct/05/smartphone-addiction-silicon-valley-dystopia

25. FS. (2017). How Filter Bubbles Distort Reality: Everything You Need to Know. https://fs.blog/2017/07/filter-bubbles/

26. Gould, W. R. (2019, October 22). Are you in a social media bubble? Here's how to tell. *NBC News*. https://www.nbcnews.com/better/lifestyle/problem-social-media-reinforcement-bubbles-what-you-can-do-about-ncna1063896

27. Nguyen, C. T. (2019, September 11). The problem of living inside echo chambers. *The Conversation*. https://theconversation.com/the-problem-of-living-inside-echo-chambers-110486

28. Grimes, D. R. (2017, December 4). Echo chambers are dangerous – we must try to break free of our online bubbles. *The Guardian*. https://www.theguardian.com/science/blog/2017/dec/04/echo-chambers-are-dangerous-we-must-try-to-break-free-of-our-online-bubbles

29. Haynes, T. (2018, May 1). Dopamine, Smartphones & You: A battle for your time. Harvard University Graduate School of Arts and Sciences. https://sitn.hms.harvard.edu/flash/2018/dopamine-smartphones-battle-time/

30. Ward, A. F., Duke, K., Gneezy, A., & Bos, M. W. (2017). Brain Drain: The Mere Presence of One's Own Smartphone Reduces Available Cognitive Capacity. *Journal of the Association for Consumer Research*, 2(2).

31. Lewis, P. (2017, October 6). 'Our minds can be hijacked': The tech insiders who fear a smartphone dystopia. *The Guardian*. https://www. theguardian.com/technology/2017/oct/05/smartphone-addiction-sili-con-valley-dystopia

32. Durov's Channel. (n.d.). https://t.me/durov/166

33. Durov's Channel. (n.d.). https://t.me/durov/166

34. Grayling, A. C. (2021). *The Frontiers of Knowledge: What We Know About Science, History and The Mind – And How We Know It.* Penguin Books.

35. Matthew, D. (2015, June 6). This is the best letter of recommendation ever. *Vox*. https://www.vox.com/2015/6/6/8738229/john-nash-recom-mendation-letter

36. Goode, E. (2015, May 24). John F. Nash Jr., Math Genius Defined by a 'Beautiful Mind', Dies at 86. *The New York Times*. https://www.nytimes. com/2015/05/25/science/john-nash-a-beautiful-mind-subject-and-nobel-winner-dies-at-86.html

37. Louis, D. (2018, May 3). Systems-Based Thinking: How Subconscious Thought Affects Medical Decision Making. *In-House*. https://in-hous-estaff.org/systems-based-thinking-subconscious-thought-affects-medi-cal-decision-making-1049

38. Synder, B. (2017, April 29). How Richard Branson and 5 other CEOs get ahead by scheduling time off. *CNBC*. https://www.cnbc. com/2017/04/28/how-richard-branson-and-5-other-ceos-get-ahead-by-scheduling-time-off.html

39. Porter, J. (2017, March 21). Why You Should Make Time for Self Reflection (Even If You Hate Doing It). *Harvard Business Review*. https://hbr. org/2017/03/why-you-should-make-time-for-self-reflection-even-if-you-hate-doing-it

40. Queen Margaret University. (n.d.). Reflection. https://www.qmu.ac.uk/media/5533/reflection-2014.pdf

41. The Peak Performance Center. (n.d.). Analytical Thinking. https://the-peakperformancecenter.com/educational-learning/thinking/types-of-thinking-2/analytical-thinking/

42. Williams, C. C., Kappen, M., Hassall, C. D., Wright, B., & Krigolson, O. E. (2019). Thinking theta and alpha: Mechanisms of intuitive and analyti-cal reasoning. *NeuroImage*, 189: 574–580. https://www.krigolsonlab.com/uploads/4/3/8/4/43848243/williams_2019.pdf

43. Devonald, E. (2019, March 27). How is this green school in Bali breaking

down the walls of learning? *Study International*. https://www.studyinter-national.com/news/how-is-this-green-school-in-bali-breaking-down-the-walls-of-learning/

44. The British Psychological Society. (2017, July). Is slowness the essence of knowledge? https://thepsychologist.bps.org.uk/volume-30/july-2017/slowness-essence-knowledge

45. Reis, R. (n.d.). Slow Knowing. Stanford University. https://tomprof.stanford.edu/posting/395

46. Boudreau, E. (2020, January 16). The Art of Slow Looking in the Classroom. Harvard Graduate School of Education. https://www.gse.harvard.edu/news/uk/20/01/art-slow-looking-classroom

47. Golden, G. (2017, December 12). Want to Make Better Decisions? Start By Slowing Down Your Thought Process. *Harvard Business Review*. https://www.inc.com/gary-golden/3-reasons-slowing-down-your-thought-process-will-help-you-make-better-decisions.html

48. Markman, A. (2020, March 15). Slow Down to Make Better Decisions in a Crisis. *Harvard Business Review*. https://hbr.org/2020/03/slow-down-to-make-better-decisions-in-a-crisis

49. Hammond, J. S., Keeney, R. L., & Raiffa, H. (1998, September–October). The Hidden Traps in Decision Making. *Harvard Business Review*. https://hbr.org/1998/09/the-hidden-traps-in-decision-making-2

50. Soll, J. B., Milkman, K. L., & Payne, J. W. (2015, May). Outsmart Your Own Biases. *Harvard Business Review*. https://hbr.org/2015/05/outsmart-your-own-biases

51. FS. (2017). Confirmation Bias And the Power of Disconfirming Evidence. https://fs.blog/2017/05/confirmation-bias/

52. Grjebine, L. (2020, October 9). Why Doubt Is Essential to Science. *Scientific American*. https://www.scientificamerican.com/article/why-doubt-is-essential-to-science/

53. Metz, C. (2016, March 16). In Two Moves, AlphaGo and Lee Sedol Redefined the Future. *Wired*. https://www.wired.com/2016/03/two-moves-alphago-lee-sedol-redefined-future/

54. Fleming, S. M. (2021, April 16). What separates humans from AI? It's doubt. *Financial Times*. https://www.ft.com/content/1ff66eb9-166f-4082-958f-debe84e92e9e

55. National Geographic. (2013, May 29). The Bigger Brains of London Taxi Drivers. https://www.nationalgeographic.com/culture/article/the-big-

ger-brains-of-london-taxi-drivers

56. Arizona State University. (2019, March 6). More than just memories: A new role for the hippocampus during learning. *Science Daily*. https://www.sciencedaily.com/releases/2019/03/190306081704.htm

57. Sarrasin, J. B., et al. (2020, May 14). Understanding Your Brain to Help You Learn Better. *Frontiers for Young Minds*. https://kids.frontiersin.org/articles/10.3389/frym.2020.00054

58. Comaford, C. (2014, November 7). The Truth About How Your Brain Gets Smarter. *Forbes*. https://www.forbes.com/sites/christine-comaford/2014/11/07/the-truth-about-how-your-brain-gets-smarter/?sh=40a3e84519bc

59. Bryan, K., & Bryan, D. (2008). Development of the Child's Brain and Behavior. In *Handbook of Clinical Child Neuropsychology*, 3rd edition. Springer.

60. Comaford, C. (2014, November 7). The Truth About How Your Brain Gets Smarter. *Forbes*. https://www.forbes.com/sites/christine-comaford/2014/11/07/the-truth-about-how-your-brain-gets-smarter/?sh=40a3e84519bc

61. Hom, M. (2015, November 30). Beyond Einstein's Brain: The Anatomy of Genius. *Elsevier SciTech Connect*. http://scitechconnect.elsevier.com/beyond-einsteins-brain-anatomy-genius/

62. Fields, R. D. (2020, March 1). The Brain Learns in Unexpected Ways. *Scientific American*. https://www.scientificamerican.com/article/the-brain-learns-in-unexpected-ways/

63. Nadella, S. (2014, February 4). Satya Nadella email to employees on first day as CEO. *Microsoft News Center*. https://news.microsoft.com/2014/02/04/satya-nadella-email-to-employees-on-first-day-as-ceo/

64. Johnson, E. (2017, May 29). Learn To Have A Brain Like Einstein's. https://www.emmajohnson.co.uk/learn-to-have-a-brain-like-einsteins/

65. Ring, K. (2016, May 2). An inside look reveals the adult brain prunes its own branches. *CIRM*. https://blog.cirm.ca.gov/2016/05/02/an-inside-look-reveals-the-adult-brain-prunes-its-own-branches/

66. Yetis, N. (2015, August 27). How Traveling Triggers Creativity. https://www.nilgunyetis.com/2015/08/27/how-traveling-triggers-creativity/

67. Mlodinow, L. (2018). *Elastic: Unlocking Your Brain's Ability to Embrace Change*. Pantheon.

68. Mlodinow, L. (2018). *Elastic: Unlocking Your Brain's Ability to Embrace Change*. Pantheon.

69. Thaler, R. H. (2021). *Nudge: The Final Edition*. Penguin Books.
70. Service, O., & Gallagher, R. (2017). *Think Small: The Surprisingly Simple Ways to Reach Big Goals*. Michael O'Mara.
71. Clear, J. (n.d.). This Coach Improved Every Tiny Thing by 1 Percent and Here's What Happened. https://jamesclear.com/marginal-gains
72. Gefter, A. (2010, January 18). Newton's apple: The real story. *New Scientist*. https://www.newscientist.com/article/2170052-newtons-apple-the-real-story/
73. Newman, T. (2017, October 25). The brain's 'daydream' network is more active than we thought. *Medical News Today*. https://www.medicalnewstoday.com/articles/319846
74. Buckner, R. L., Andrews-Hanna, J. R., & Schacter, D. L. (2008). The brain's default network: Anatomy, function, and relevance to disease. The year in cognitive neuroscience 2008. Blackwell Publishing.
75. Mlodinow, L. (2018). *Elastic: Unlocking Your Brain's Ability to Embrace Change*. Pantheon.
76. PCA. Creating your own reality: The power of imagination. https://pca-global.com/creating-reality-power-imagination/
77. Mlodinow, L. (2018). *Elastic: Unlocking Your Brain's Ability to Embrace Change*. Pantheon.
78. Dietrich, A. (2004). Neurocognitive mechanisms underlying the experience of flow. *Conscious Cogn*. https://pubmed.ncbi.nlm.nih.gov/15522630/
79. Schootstra, E., Deichmann, D., & Dolgova, E. (2017, August 29). Can 10 Minutes of Meditation Make You More Creative. *Harvard Business Review*. https://hbr.org/2017/08/can-10-minutes-of-meditation-make-you-more-creative
80. Zhang, Z., Luh, WM., Duan, W., Zhou, G. D., Weinschenk, G., Anderson, A. K., & Dai, W. (2021, May 31). Longitudinal effects of meditation on brain resting-state functional connectivity. *Scientific Reports*. https://www.nature.com/articles/s41598-021-90729-y
81. Buddha Weekly. (n.d.). Cankama Sutta: Walking Meditation Sutra: Put some mileage on your Buddhist practice with formal mindful walking. https://buddhaweekly.com/cankama-sutta-walking-meditation-sutra-put-mileage-buddhist-practice-formal-mindful-walking/
82. Mlodinow, L. (2018). *Elastic: Unlocking Your Brain's Ability to Embrace Change*. Pantheon.

83. Lewicki, P., Hoffman, H., & Czyzewska, M. (1987). Unconscious Acquisition of Complex Procedural Knowledge. *Journal of Experimental Psychology: Learning, Memory & Cognition.* http://www.mwbp.org/research/lewicki/Lewicki%20(1987).pdf

84. Mlodinow, L. (2018). *Elastic: Unlocking Your Brain's Ability to Embrace Change.* Pantheon.

85. Boaler, J. (2019, October 28). Why Struggle Is Essential for the Brain – and Our Lives. *EdSurge.* https://www.edsurge.com/news/2019-10-28-why-struggle-is-essential-for-the-brain-and-our-lives

86. Mlodinow, L. (2018). *Elastic: Unlocking Your Brain's Ability to Embrace Change.* Pantheon.

87. Syed, M. (2015, November 14). Viewpoint: How creativity is helped by failure. *BBC.* https://www.bbc.com/news/magazine-34775411

88. Mlodinow, L. (2018). *Elastic: Unlocking Your Brain's Ability to Embrace Change.* Pantheon.

89. Schoemaker, P. J. H., & Gunther, R. E. (2006, June). The Wisdom of Deliberate Mistakes. *Harvard Business Review.* https://hbr.org/2006/06/the-wisdom-of-deliberate-mistakes

90. Koch, C. (2016, March 19). How the Computer Beat the Go Master. *Scientific American.* https://www.scientificamerican.com/article/how-the-computer-beat-the-go-master/

91. Kehoe, R. (2020, September 10). A secret of science: Mistakes boost learning. *Science News for Students.* https://www.sciencenewsforstudents.org/article/secret-science-mistakes-boost-understanding

92. Kehoe, R. (2020, September 10). A secret of science: Mistakes boost learning. *Science News for Students.*

93. Grant, A. (2021). *Think Again: The Power of Knowing What You Don't Know.* Viking.

94. Tenney, E. R., Costa, E., & Watson, R. M. (2021, June 16). Why Business Schools Need To Teach Experimentation. *Harvard Business Review.* https://hbr.org/2021/06/why-business-schools-need-to-teach-experimentation

95. FS. (2017). Thought Experiment: How Einstein Solved Difficult Problems. https://fs.blog/2017/06/thought-experiment/

96. Fleming, S. M. (2021, April 16). What separates humans from AI? It's doubt. *Financial Times.* https://www.ft.com/content/1ff66eb9-166f-4082-958f-debe84e92e9e

97. Mobley, M. (2018, August 14). Rohit Bhargava: A 'near-futurist' scours data for hidden clues about how the world works. Microsoft. https://news.microsoft.com/stories/people/rohit-bhargava.html

98. Yale School of Management. (2019, February 13). The Seduction of Mistakes. https://som.yale.edu/blog/the-seduction-of-mistakes

99. Ozenc, K. (2016, April 2). Introducing Ritual Design: meaning, purpose and behavior change. *Ritual Design Lab.* https://medium.com/ritual-design/introducing-ritual-design-meaning-purpose-and-behavior-change-44d26d484edf

100. Quandt, K. R. (n.d.). What happens in the brain when you're alone. *Headspace.* https://www.headspace.com/articles/brain-alone

101. Guth, R. A. (2008, March 28). In Secret Hideaway, Bill Gates Ponders Microsoft's Future. *The Wall Street Journal.* https://www.wsj.com/articles/SB111196625830690477

102. Clifford, C. (2019, July 28). Bill Gates took solo 'think weeks' in a cabin in the woods – why it's a great strategy. *CNBC.* https://www.cnbc.com/2019/07/26/bill-gates-took-solo-think-weeks-in-a-cabin-in-the-woods.html

103. Clifford, C. (2019, July 28). Bill Gates took solo 'think weeks' in a cabin in the woods – why it's a great strategy. *CNBC.* https://www.cnbc.com/2019/07/26/bill-gates-took-solo-think-weeks-in-a-cabin-in-the-woods.html

104. Jabr, F. (2016, September 1). Q&A: Why a Rested Brain Is More Creative. *Scientific American.* https://www.scientificamerican.com/article/q-a-why-a-rested-brain-is-more-creative/

105. MacKay, J. (2016, June 28). Is Solitude the Secret to Unlocking Our Creativity. *Observer.* https://observer.com/2016/06/is-solitude-the-secret-to-unlocking-our-creativity/

106. Rushing, S. (2016, January 25). 25 Daily Rituals of History's Most Successful People. *Observer.* https://observer.com/2016/01/25-daily-rituals-of-historys-most-successful-people/

107. Oppong, T. (2017, February 7). Daily routines of Nikola Tesla, Mozart, Hemingway, Woody Allen, Maya Angelou, van Gogh, Stephen King, and Nabokov. *CNBC.* https://www.cnbc.com/2017/02/07/daily-routines-of-tesla-mozart-hemingway-and-more.html

108. Schools, D. (2017, March 6). Exactly How Much Sleep Mark Zuckerberg, Jack Dorsey, and Other Successful Business Leaders Get. *Inc.* https://

www.inc.com/dave-schools/exactly-how-much-sleep-mark-zuckerberg-jack-dorsey-and-other-successful-business.html

109. Kluger, J. (2017, April 30). How to Wake Up To Your Creativity. *Time.* https://time.com/4737596/sleep-brain-creativity/

110. National Geographic. (2020). *Sleep.*

111. Kluger, J. (2017, April 30). How to Wake Up To Your Creativity. Time. https://time.com/4737596/sleep-brain-creativity/

112. Yong, E. (2018, May 15). A New Theory Linking Sleep and Creativity. *The Atlantic.* https://www.theatlantic.com/science/archive/2018/05/sleep-creativity-theory/560399/

113. Pang, A. S.-K. (2017, November 7). Winston Churchill's Secret Productivity Weapon. https://michaelhyatt.com/naps/

114. Weir, K. (2016). The science of naps. American Psychological Association. https://www.apa.org/monitor/2016/07-08/naps

115. Sample, I. (2010, February 22). People learn more after a siesta, say scientists. *The Guardian.* https://www.theguardian.com/science/2010/feb/21/naps-improve-learning-ability

116. Yeager, A. (2018, November 1). How Exercise Reprograms Our Brains. *The Scientist.* https://www.the-scientist.com/features/this-is-your-brain-on-exercise-64934

117. Harvard Health Publishing. (2021, February 15). Exercise can boost your memory and thinking skills. https://www.health.harvard.edu/mind-and-mood/exercise-can-boost-your-memory-and-thinking-skills

118. Wood, J. (2021, February 23) Exercise not only helps with mental health – it makes us more creative too, say scientists. World Economic Forum. https://www.weforum.org/agenda/2021/02/exercise-mental-health-creativity/

119. Reynolds, G. (2016, December 14). Running as the Thinking Person's Sport. *The New York Times.* https://www.nytimes.com/2016/12/14/well/move/running-as-the-thinking-persons-sport.html

120. Klemm, W. R. (2013, March 14). Why Writing by Hand Could Make You Smarter. *Psychology Today.* https://www.psychologytoday.com/us/blog/memory-medic/201303/why-writing-hand-could-make-you-smarter

121. University of Tokyo. (2021, March 19). Stronger Brain Activity After Writing on Paper Than on Tablet or Smartphone. *NeuroScience News.* https://neurosciencenews.com/hand-writing-brain-activity-18069/

122. Fernyhough, E. (2021, January 19). How Writing Changes Your Brain. *The Brave Writer*. https://medium.com/the-brave-writer/how-writing-changes-your-brain-75c9087f37e7

123. McSpadden, K. (2015, May 14). You Now Have a Shorter Attention Span Than a Goldfish. *Time*. https://time.com/3858309/attention-spans-goldfish/

124. McSpadden, K. (2015, May 14). You Now Have a Shorter Attention Span Than a Goldfish. *Time*. https://time.com/3858309/attention-spans-goldfish/

125. MacKay, J. (2017, September 28). Science Says These 7 Attention Exercises Will Instantly Make You More Focused. *Inc*. https://www.inc.com/jory-mackay/sciences-says-these-7-attention-exercises-will-make-you-more-focused-right-now.html

126. Koetsier, J. (2020, September 26). Global Online Content Consumption Doubled in 2020. *Forbes*. https://www.forbes.com/sites/johnkoetsier/2020/09/26/global-online-content-consumption-doubled-in-2020/?sh=1d2f810c2fde

127. Jabr, F. (2013, April 11). The Reading Brain in the Digital Age: The Science of Paper versus Screens. *Scientific American*. https://www.scientificamerican.com/article/reading-paper-screens/

128. Leroy, S. (n.d.). Attention Residue. University of Washington. https://www.uwb.edu/business/faculty/sophie-leroy/attention-residue

129. American Psychological Association. (2006, March 20). Multitasking: Switching costs. https://www.apa.org/research/action/multitask

130. Dore, M. (2020, February 1). How to reduce 'attention residue' in your life. *BBC*. https://www.bbc.com/worklife/article/20200130-the-life-hack-to-reduce-admin-and-carve-out-downtime

131. Wiginton, K. (2021, June 1). Your ability to focus may be limited to 4 or 5 hours a day. Here's how to make the most of them. *The Washington Post*. https://www.washingtonpost.com/lifestyle/wellness/productivity-focus-work-tips/2021/05/31/07453934-bfd0-11eb-b26e-53663e6be6ff_story.html

132. Thornton, B., Faires, A., Robbins, M., & Rollins, E. (2014, January 1). The Mere Presence of a Cell Phone May be Distracting. Hogrefe Econtent. https://econtent.hogrefe.com/doi/10.1027/1864-9335/a000216

133. Duke, K., Ward, A., Gneezy, A., & Bos, M. (2018, March 20). Having Your Smartphone Nearby Takes a Toll on Your Thinking. *Harvard Busi-*

ness Review. https://hbr.org/2018/03/having-your-smartphone-nearby-takes-a-toll-on-your-thinking

134. McCombs School of Business. (2018, February 21). The Effects of Smartphones on Studying. University of Texas. https://research.utexas.edu/showcase/articles/view/the-effects-of-smartphones-on-studying

135. Duke, K., Ward, A., Gneezy, A., & Bos, M. (2018, March 20). Having Your Smartphone Nearby Takes a Toll on Your Thinking. *Harvard Business Review.* https://hbr.org/2018/03/having-your-smartphone-nearby-takes-a-toll-on-your-thinking

136. Carleton, R. N. (2016). Fear of the unknown: One fear to rule them all? *Journal of Anxiety Disorders,* 41. https://www.sciencedirect.com/science/article/pii/S0887618516300469

137. Glaser, J. E. (2013, February 28). Your Brain Is Hooked on Being Right. *Harvard Business Review.* https://hbr.org/2013/02/break-your-addiction-to-being

138. Walsh, R. (2017, November 1). What to Do When Work Stress (Literally) Makes You Sick. *Harvard Business Review.* https://hbr.org/2017/11/what-to-do-when-work-stress-literally-makes-you-sick

139. American Psychological Association. (2017, July). What is Exposure Therapy. https://www.apa.org/ptsd-guideline/patients-and-families/exposure-therapy

140. Stanborough, R. J. (2019, December 18). What Are Cognitive Distortions and How Can You Change These Thinking Patterns. *Healthline.* https://www.healthline.com/health/cognitive-distortions

141. Chamorro-Premuzic, T. (2015, March 25). Why Group Brainstorming Is a Waste of Time. *Harvard Business Review.* https://hbr.org/2015/03/why-group-brainstorming-is-a-waste-of-time

142. Interaction Design Foundation. (n.d.). Design Thinking. https://www.interaction-design.org/literature/topics/design-thinking

143. Wedell-Wedellsborg, T. (2017). Are You Solving the Right Problems. *Harvard Business Review.* https://hbr.org/2017/01/are-you-solving-the-right-problems

144. IBM Enterprise Design Thinking. (n.d.). Empathy Map. https://www.ibm.com/design/thinking/page/toolkit/activity/empathy-map

145. Centeno, R. (2020). Effect of Mindfulness on Empathy and Self-Compassion: An Adapted MBCT Program on Filipino College Students. *Behavioral* (Basel), 10(3): 61. https://doi.org/10.3390/bs10030061

146. Au, I. (2015, January 21). Mindfulness Practices For Greater Focus, Empathy, and Creativity. *Design Your Life*. https://medium.com/design-your-life/mindfulness-practices-for-design-9f8b7f1af047

147. IBM Enterprise Design Thinking. (n.d.). The Loop drives us. https://www.ibm.com/design/thinking/page/framework/loop

148. FS. (2013, October). Inversion and The Power of Avoiding Stupidity. https://fs.blog/2013/10/inversion/

149. Chakraborty, A. (n.d.). Inversion: Working Backwards to Solve Problems. *Coffee and Junk*. https://coffeeandjunk.com/inversion/

150. Chakraborty, A. (n.d.). Inversion: Working Backwards to Solve Problems. *Coffee and Junk*. https://coffeeandjunk.com/inversion/

151. Bariso, J. (2019, December 19). Amazon Has a Secret Weapon Known as 'Working Backwards' – and It Will Transform the Way You Work. *Inc.* https://www.inc.com/justin-bariso/amazon-uses-a-secret-process-for-launching-new-ideas-and-it-can-transform-way-you-work.html

152. Pratte, R. (2021, April 7). Downstream Benefits From PR/FAQ, Working Backwards. *Innovation Catalyst*. https://innovationcatalyst.io/home/downstream-benefits-from-prfaq-working-backwards

153. Bryar, C., & Carr, B. (2021, February 9). *Working Backwards: Insights, Stories, and Secrets from Inside Amazon*. St Martin's Press.

154. Greene, B. (2013, July). How the Higgs Boson Was Found. *Smithsonian Magazine*. https://www.smithsonianmag.com/science-nature/how-the-higgs-boson-was-found-4723520/

155. Gates, B. (2020, September 1). We're finally learning why countries excel at saving lives. *Gates Notes*. https://www.gatesnotes.com/Health/Exemplars-in-Global-Health?WT.mc_id=2020090113000

156. Waters, M. (2020, September 12). The company that wants to preserve our data for 500+ years. *The Hustle*. https://thehustle.co/arctic-world-archive-svalbard/

157. ICMA. Positive Problem Solving: How Appreciative Inquiry Works. *InFocus*, 43(3). https://www.sog.unc.edu/sites/www.sog.unc.edu/files/course_materials/Positive%20Problem%20Solving%20ICMA%20Press.pdf

158. David Cooperrider. (n.d.). What is Appreciative Inquiry. https://www.davidcooperrider.com/ai-process/

159. FS. (2018, April). First Principles: The Building Blocks of True Knowledge. https://fs.blog/2018/04/first-principles/

160. FS. (n.d.). Who Is Richard Feynman? The Curious Character Who Mastered Thinking and Physics. https://fs.blog/intellectual-giants/richard-feynman/

161. FS. (2021, February). The Feynman Learning Technique. https://fs.blog/2021/02/feynman-learning-technique/

162. FS. (2018, April). First Principles: The Building Blocks of True Knowledge. https://fs.blog/2018/04/first-principles/

163. Amos, J. (2021, May 6). SpaceX Starship prototype makes clean landing. *BBC*. https://www.bbc.com/news/science-environment-57004604

164. FS. (2018, April). First Principles: The Building Blocks of True Knowledge. https://fs.blog/2018/04/first-principles/

165. Ruggiero, N. (2020, July 8). Why Singapore Airlines Was Voted the Best International Airline 25 Years in a Row. *Travel + Leisure*. https://www.travelandleisure.com/worlds-best/singapore-airlines-best-international-airline-25-years

166. Wirtz, J. (2019, October 17). How do innovators stay innovative? Here's a look at how SIA does it. *Today*. https://www.todayonline.com/commentary/how-do-innovators-stay-innovative

167. Tikkamaki, K., Heikkila, P., and Ainasoja, M. (2016). Positive Stress and Reflective Practice Among Entrepreneurs. *J Entrepreneurship Manag Innov*, 2(1): 35–56.

168. Kaimal, G., Ray, K., and Muniz, J. (2016) Reduction of Cortisol Levels and Participants' Responses Following Art Making. *Art Ther (Alex)*, 33(2): 74–80.

169. Thomée, S., Härenstam, A., and Hagberg, M. (2011). Mobile phone use and stress, sleep disturbances, and symptoms of depression among young adults: A prospective cohort study. *BMC Public Health*, 11: 66.

170. Mineo, Liz. (2017, April 11). Harvard study, almost 80 years old, has proved that embracing community helps us live longer, and be happier. *The Harvard Gazette*. https://news.harvard.edu/gazette/story/2017/04/over-nearly-80-years-harvard-study-has-been-showing-how-to-live-a-healthy-and-happy-life/

About the Author

Dr Andreas Raharso is former Dean of the Hay Group Global Research Centre for Strategy Execution, which was set up in affiliation with Harvard University and supported by the Singapore Government. A thought-leader in disruptive innovation, he created the original People Analytics module at INSEAD Business School, and two postgraduate modules – Next Practice and People Strategy – at the National University of Singapore (NUS) Business School. Andreas has also collaborated with leading businesses such as Microsoft, Amazon, Google, Alibaba, Airbnb, GE, P&G and Gojek. He holds degrees in Economics, Finance and Data Science, and a PhD in Management.